Electronic Waste Management

ISSUES IN ENVIRONMENTAL SCIENCE AND TECHNOLOGY

EDITORS:

R.E. Hester, University of York, UK
R.M. Harrison, University of Birmingham, UK

EDITORIAL ADVISORY BOARD:

TITLES IN THE SERIES:

How to obtain future titles on publication

A subscription is available for this series. This will bring delivery of each new volume immediately on publication and also provide you with online access to each title via the Internet. For further information visit http://www.rsc.org/Publishing/Books/issues or write to the address below.

For further information please contact:
Sales and Customer Care, Royal Society of Chemistry, Thomas Graham House, Science Park, Milton Road, Cambridge, CB4 0WF, UK
Telephone: +44 (0)1223 432360, Fax: +44 (0)1223 426017, Email: sales@rsc.org

ISSUES IN ENVIRONMENTAL SCIENCE AND TECHNOLOGY

EDITORS: R.E. HESTER AND R.M. HARRISON

27
Electronic Waste Management

RSCPublishing

ISBN: 978-0-85404-112-1
ISSN: 1350-7583

A catalogue record for this book is available from the British Library

Published by The Royal Society of Chemistry,
Thomas Graham House, Science Park, Milton Road,
Cambridge CB4 0WF, UK

Registered Charity Number 207890

For further information see our website at www.rsc.org

Preface

The accelerating pace of consumption of materials, energy and other resources needed to maintain the production of electrical and electronic products in the developed and developing world is clearly unsustainable. Recent growth in the use of mobile telephones, personal computers and flat screen TVs is spectacular. Paralleling this growth is a progressive reduction in the lifetimes of these and other consumer electronics, often due to reasons of fashion rather than function, resulting in a massive increase in the volumes of waste generated. It has been recognised that much higher levels of recycling and recovery of the materials used in the manufacture of electrical and electronic products, whether metals or speciality organic chemicals and plastics, has become essential and new legislation has been introduced world-wide to promote this.

Given the technical complexity of much of the so-called eWaste and the difficulties that this presents for recovery and re-use of precious materials, much attention is being given to the design of electrical and electronic products in order to facilitate their treatment at end of life. Cradle-to-grave planning considerations at the design stage are being supplanted by cradle-to-cradle ones, driven by both environmental and economic factors and by legislation such as the WEEE, RoHS, EuP and REACH Directives recently imposed in the European Community.

In this book we have brought together the collective expertise of a group of leading practitioners in the field of electrical and electronic waste management. Martin Goosey, of Loughborough University, has written an introduction and overview for the opening chapter wherein he addresses the issues of sustainability and alternatives to dumping this waste in third world and Far Eastern countries. Producer Responsibility legislation generated by the European Commission is examined alongside other global regulation of the industry, such as that applied in Japan where much of the world's consumer electronics originate. The scope of this first chapter, from household appliances such as refrigerators and washing machines to mobile phones and flat-screen televisions, reflects that of the book as a whole.

Issues in Environmental Science and Technology, 27
Electronic Waste Management
Edited by R.E. Hester and R.M. Harrison
© Royal Society of Chemistry 2009
Published by the Royal Society of Chemistry, www.rsc.org

The materials used in manufacturing electrical and electronic goods form the subject of chapter 2, written jointly by Gary Stevens of the University of Surrey and Martin Goosey. The recent legislative constraints on the use of materials such as heavy metals and certain brominated organic chemicals commonly used as flame retardants, are examined and alternative materials such as those used to replace lead solders are discussed. Laminate materials used for printed circuit boards, indium tin oxide thin films used in LCD displays, and the wide variety of engineering thermoplastics used in applications as diverse as car instrument panels and medical electronics are other examples of materials reviewed here.

In chapter 3, Ian Holmes of C-Tech Innovations Ltd surveys the well-established methods of dealing with waste by dumping in landfill sites or by incineration. Although current thinking is that landfilling is the last resort for waste disposal, much of the electrical and electronic waste generated still is disposed in this way along with other municipal solid waste, sometimes after shipping abroad. Recycling and recovery of eWaste is reviewed by Darren Kell, also of C-Tech, in chapter 4, while an integrated approach to recycling is described in chapter 5 by Rod Kellner of Kellner Environmental Ltd. These chapters cover the many technical aspects of recycling and encompass the full range of products, materials and treatments. The work of the European Recycling Platform, which is a pan-European Producer Compliance Scheme involving several major companies, is described by Scott Butler of ERP Recycling in chapter 6.

In chapter 7, Avtar Matharu of the University of York and Yanbing Yu of the Beijing Optoelectronics Technology Co Ltd present a detailed review of liquid crystal displays, from devices to recycling, while in chapter 8 Mark Dempsey and Kirstie McIntyre of Hewlett Packard examine different models for eWaste management from around the world from an extended producer responsibility perspective. And finally, in chapter 9, Patrick Baird, Henryk Herman and Gary Stevens of GnoSys UK at the University of Surrey describe their work on the development of a range of spectroscopic methods for the rapid identification and evaluation of electronics enclosure plastics and their additives, such as flame retardants.

Overall, the volume provides a comprehensive overview of current issues in the fast-moving field of eWaste management which we believe will be of immediate value not only to engineers and managers in this important industry but also to environmentalists and policymakers, as well as to students in environmental science and engineering and management courses.

Ronald E Hester
Roy M Harrison

Contents

Issues in Environmental Science and Technology, 27
Electronic Waste Management
Edited by R.E. Hester and R.M. Harrison
© Royal Society of Chemistry 2009
Published by the Royal Society of Chemistry, www.rsc.org

Chapter 2 Materials Used in Manufacturing Electrical and Electronic Products
Gary C. Stevens and Martin Goosey

Chapter 3 Dumping, Burning and Landfill
Ian Holmes

Chapter 4 Recycling and Recovery
Darren Kell

Chapter 5 Integrated Approach to e-Waste Recycling
Rod Kellner

Chapter 6 European Recycling Platform (ERP): a Pan-European Solution to WEEE Compliance
Scott Butler

**Chapter 8   The Role of Collective versus Individual Producer
            Responsibility in e-Waste Management: Key Learnings
            from Around the World**
Mark Dempsey and Kirstie McIntyre

Chapter 9   Rapid Assessment of Electronics Enclosure Plastics
Patrick J. Baird, Henryk Herman and Gary C. Stevens

Editors

Ronald E. Hester, BSc, DSc(London), PhD(Cornell), FRSC, CChem

Ronald E. Hester is now Emeritus Professor of Chemistry in the University of York. He was for short periods a research fellow in Cambridge and an assistant professor at Cornell before being appointed to a lectureship in chemistry in York in 1965. He was a full professor in York from 1983 to 2001. His more than 300 publications are mainly in the area of vibrational spectroscopy, latterly focusing on time-resolved studies of photoreaction intermediates and on bio-molecular systems in solution. He is active in environmental chemistry and is a founder member and former chairman of the Environment Group of the Royal Society of Chemistry and editor of 'Industry and the Environment in Perspective' (RSC, 1983) and 'Understanding Our Environment' (RSC, 1986). As a member of the Council of the UK Science and Engineering Research Council and several of its sub-committees, panels and boards, he has been heavily involved in national science policy and administration. He was, from 1991 to 1993, a member of the UK Department of the Environment Advisory Committee on Hazardous Substances and from 1995 to 2000 was a member of the Publications and Information Board of the Royal Society of Chemistry.

Roy M. Harrison, BSc, PhD, DSc(Birmingham), FRSC, CChem, FRMetS, Hon MFPH, Hon FFOM

Roy M. Harrison is Queen Elizabeth II Birmingham Centenary Professor of Environmental Health in the University of Birmingham. He was previously Lecturer in Environmental Sciences at the University of Lancaster and Reader and Director of the Institute of Aerosol Science at the University of Essex. His more than 300 publications are mainly in the field of environmental chemistry, although his current work includes studies of human health impacts of atmospheric pollutants as well as research into the chemistry of pollution phenomena. He is a past Chairman of the Environment Group of the Royal Society of Chemistry for whom he has edited 'Pollution: Causes, Effects and Control' (RSC, 1983; Fourth Edition, 2001) and 'Understanding our Environment: An Introduction to Environmental Chemistry and Pollution'

(RSC, Third Edition, 1999). He has a close interest in scientific and policy aspects of air pollution, having been Chairman of the Department of Environment Quality of Urban Air Review Group and the DETR Atmospheric Particles Expert Group. He is currently a member of the DEFRA Air Quality Expert Group, the DEFRA Expert Panel on Air Quality Standards, and the Department of Health Committee on the Medical Effects of Air Pollutants.

List of Contributors

Patrick Baird, *GnoSys UK Ltd, University of Surrey, Guildford, Surrey GU2 7XH, UK*

Scott Butler, *ERP UK Ltd, Mazars, Tower Bridge House, St Katharine's Way, London, E1W 1DD, UK*

Mark Dempsey, *Hewlett-Packard, Cain Road, Bracknell, RG12 1HN, UK*

Martin Goosey, *Wolfson School of Mechanical and Manufacturing Engineering, Loughborough University, Ashby Road, Loughborough, LE11 3TU, UK*

Henryk Herman, *GnoSys UK Ltd, University of Surrey, Guildford, Surrey GU2 7XH, UK*

Ian Holmes, *C-Tech Innovation Ltd, Capenhurst Technology Park, Capenhurst, Cheshire, CH1 6EH, UK*

Darren Kell, *C-Tech Innovation Ltd, Capenhurst Technology Park, Capenhurst, Cheshire, CH1 6EH, UK*

Rod Kellner, *P K Kellner Environmental Ltd, Edmund House, Coventry Road, Pailton, Warwickshire, CV23 0QB, UK*

Avtar Matharu, *Department of Chemistry, University of York, York, England YO10 5DD, UK*

Kirstie McIntyre, *Hewlett-Packard, Cain Road, Bracknell, RG12 1HN, UK*

Gary Stevens, *GnoSys UK Ltd, University of Surrey, Guildford, Surrey GU2 7XH, UK*

Yanbing Wu, *Beijing BOE Optoelectronics Technology Co. Ltd, Xihuanzhang Road #8, BDA, 100176, Beijing, P.R. China, currently at the University of York*

Introduction and Overview

MARTIN GOOSEY

1 Introduction

In recent years there has been growing concern about the negative impacts that industry and its products are having on both society and the environment in which we live. The concept of sustainability and the need to behave in a more sustainable manner has therefore received increasing attention. With the world's population growing rapidly and generally improving wealth, the consumption of materials, energy and other resources has been accelerating in a way that cannot be sustained. With issues such as global warming also now more openly acknowledged as being significantly influenced by our activities, there is a clear need to address the way society uses, and often wastes, valuable resources. In short, we have to behave more sustainably. There are a number of useful definitions of sustainability and the World Commission on Environment & Development has defined it as:

'Meeting the needs of the present generations without compromising the ability of future generations to meet their own needs'

This is a good top-level definition but, in the context of industry, it needs to be more specifically focused to encompass the typical requirements of businesses and a more appropriate definition is:

'Adopting strategies and activities that meet the needs of the enterprise and its stakeholders today while protecting, sustaining and enhancing the human and natural resources that will be needed in the future'

Issues in Environmental Science and Technology, 27
Electronic Waste Management
Edited by R.E. Hester and R.M. Harrison
© Royal Society of Chemistry 2009
Published by the Royal Society of Chemistry, www.rsc.org

One area in which there has been much concern about the lack of sustainable behaviour is in the manufacture, use and disposal of electrical and electronic products. The electronics industry provides us with the devices that have become so essential to our modern way of life and yet it also represents an area where the opportunities to operate in a sustainable way have not yet been properly realised. In fact, much electrical and electronic equipment (EEE) is typically characterised by a number of factors, including improved performance and reduced cost in each new generation of product, that actually encourage unsustainable behaviour. Products such as mobile phones are often treated as fashion items and are replaced long before their design lifetimes have expired; see Figure 1.

With products increasingly having short lifecycles, using hazardous materials and processes, and generating waste both during manufacture and at end of life, the manufacturers of EEE have become an increasingly popular and easy target for environmental groups such as Greenpeace, who have embarrassingly highlighted the deficiencies of many large international electronics companies.[1] There has also been much recent negative publicity for manufacturers about the eventual fate of their products at end of life and the effective dumping of electrical and electronic waste in Third World and Far Eastern countries.[2] Clearly, while western society has demonstrated that it is keen to embrace the benefits that modern electrical and electronic products can bring, when it comes to end of life and disposal, we have been happy to allow other parts of the world to deal with the problem.

Figure 1   End-of-life mobile telephones.

In an acknowledgement that society in general, and the electrical/electronics industry in particular, needs to operate in a more responsible and sustainable manner, the European Commission (EC) has, over the last few years, introduced a suite of Producer Responsibility legislation to address the problem. This is being driven by the EC to achieve a number of objectives aimed at a more sustainable approach to resource use and a reduction in the quantity of waste going to landfill. It also aims to divert end-of-life products for re-use, recycling and other forms of recovery, as well as proscribing the use of certain hazardous materials and reducing energy consumption through the product lifecycle. Interestingly, Producer Responsibility is an extension of the 'polluter pays' principle and it places responsibility for end-of-life management on the original producer. In summary, Producer Responsibility legislation aims to encourage producers to design, manufacture and market products that:

- reduce or eliminate the use of hazardous materials
- use greater amounts of recyclate
- can be more easily treated at end of life
- minimise waste
- can be re-used
- use fewer resources throughout their life

Within Europe, there are numerous Directives and Regulations aimed at implementing Producer Responsibility and key examples important to the electrical and electronics industries include the WEEE, RoHS and Energy-using Products Directives, as well as the REACH Regulations.[3]

There is clearly a need for the electronics industry to operate in a more sustainable manner, both to meet the requirements of the increasingly stringent legislation and to satisfy the needs of customers who also expect industry to have high environmental standards. The electronics industry can achieve these aims through the adoption of new manufacturing processes, the use of new materials and the development of enhanced recovery and re-use strategies at end of life. While this can already often be achieved by industry itself, there are also longer-term opportunities that will only be addressed *via* further research and development.

This opening chapter gives a broad introduction to the issues of sustainability within the context of end-of-life electrical and electronic products. The following text seeks to outline the nature of electrical and electronic equipment waste, the scale of the problem and current practices to deal with it. The way that Waste Electrical and Electronic Equipment (WEEE) has been, and continues to be, treated is described and details of new, more sustainable approaches to waste treatment are outlined. It is clear that EEE needs to be considered in a more holistic way, with a 'cradle to cradle' rather than 'cradle to grave' approach. Recent Producer Responsibility legislation, largely led by Europe, has set the future agenda and, globally, there is now an acknowledgment both of the scale of the problem and of the need for innovative solutions.

2   WEEE – The Scale of the Problem

WEEE has been Europe's fastest-growing waste stream for a number of years and it has been estimated that an average UK citizen born in 2003 will be responsible for generating around 8 tonnes of WEEE during her or his lifetime. The quantities of WEEE produced are both very large and growing. For example, the total amount of European WEEE produced in 1998 was estimated as being 6 million tonnes, with the figure having grown to between 8.3 and 9.1 million tonnes by 2005. For the period covering the next 12 years, it has been predicted that total European WEEE arisings will grow annually at between 2.5% and 2.7% to reach a figure in excess of 12 million tonnes by 2020. (Although this increase does partly represent a real growth in the quantities of WEEE that Europe generates, it should also be remembered that Europe has also grown in size to embrace a number of new member states.)

In recent years the Royal Society for the encouragement of Arts, Manufactures and Commerce (RSA) has highlighted the large volume of WEEE that each person is responsible for generating during their lifetimes *via* the construction of the 'RSA WEEE Man' shown in Figure 2. Designed by Paul Bonomini, the 'RSA WEEE Man' is a huge robotic figure made of scrap electrical and electronic equipment. Weighing 3.3 tonnes and standing 7 metres tall, it represents the average amount of electrical and electronic products each of us throws away during our lifetime.

In the UK, almost 2 million tonnes of WEEE are generated each year. Data compiled in earlier studies on arisings of WEEE, expressed as weight and units for the categories defined by the WEEE Directive, used sales data from 2003 as the starting point. Information was obtained from manufacturers, retailers, trade associations and market research organisations. The studies estimated that 939,000 tonnes of domestic equipment were discarded in the UK in 2003 and this comprised 93 million items of equipment. Table 1 shows the arisings of domestic WEEE in the UK in 2003. (No information on medical devices and automatic dispensers was obtained and therefore is not included in the table.)

Clearly, WEEE represents a serious problem, not just in terms of how its treatment and disposal is ultimately managed but also in the broader context of sustainability and the waste of valuable and finite resources.

3   Legislative Influences on Electronics Recycling

3.1   *Producer Responsibility Legislation*

Following acknowledgment that the volumes of WEEE arising in the European Union were very large and increasing year on year, the EC introduced a range of legislation aimed directly at tackling the problem. The two key, and perhaps best known, pieces of legislation are the WEEE and RoHS Directives. After over 10 years of debate, these Directives have now become a reality and they have had a significant impact on the way manufacturers design, produce and dispose of their products. The WEEE Directive, however, is just one part of a

Figure 2   The RSA WEEE Man; an RSA environmental awareness initiative to highlight the growing problem of WEEE in the UK and across Europe. (Photograph by David Ramkalawon.)

much larger policy mechanism within the EC that is aimed at introducing Producer Responsibility. This makes the producers (in this case, of electrical and electronic equipment) legally responsible for the recovery and recycling of their products when they are finally disposed of at end of life. In addition to these recently implemented directives, there are also a number of other pieces of pending legislation that will have at least some impact on aspects of electronic waste management. Key examples here include the Energy-using Products Directive and the new chemicals legislation known as the REACH Regulations.

Table 1   Domestic WEEE in the UK for 2003.

Categories of domestic WEEE	Tonnage discarded (k tonnes)	%	Units discarded (millions)	%
Large household appliances	644	69	14	16
Small household appliances	80	8	30	31
IT/Telecoms equipment	68	7	21	23
Consumer equipment	120	13	12	13
Tools	23	2	5	5
Toys, leisure and sports equipment	2	>1	2	2
Lighting	2	>1	9	10
Monitoring and control equipment	>1	>1	>1	>1
Total	**940**	**100**	**93**	**100**

3.2   The WEEE Directive[4,5]

The Waste Electrical and Electronic Equipment (WEEE) Directive directly controls the disposal of end-of-life equipment and the percentage going to landfill, as well as setting targets for the percentages of a product that have to be recovered and recycled. The WEEE Directive specifies ten categories of types of electrical and electronic equipment and each category has a defined recycling and recovery target. All recycling and recovery targets are based on a percentage of total product weight. Although there is a huge amount of specific detail within the WEEE Directive, its broad aim is to reduce the volume of electrical and electronic waste consigned to landfill, increase the recovery and recycling of electrical and electronic waste and minimise the lifecycle environmental impact of the electrical and electronic equipment sector.

The basic aims of the WEEE Directive can be summarised as follows:

- Separate collection of WEEE (4 kg per head of population)
- Treatment according to agreed standards
- Recovery and recycling to meet set targets
- Producer pays from collection onwards (retail)
- Option for business users to pay some or all of costs
- Retailers to offer take-back of end-of-life equipment
- Consumers to return WEEE free of charge

By introducing guidelines and requirements such as the provision of information for recycling and the design of products to aid re-use, recovery and recycling, the WEEE Directive aims to improve the environmental performance of all operators involved in the lifecycle of EEE, *i.e.* producers, customers and recyclers.

3.3 The RoHS Directive[6,7]

The 'Restriction of the use of certain Hazardous Substances in electrical and electronic equipment' (RoHS) Directive was originally contained within the text of the WEEE Directive, but it has subsequently been removed and now exists as a stand-alone Directive that complements the WEEE Directive. The key objective of the RoHS Directive is the protection of human health and the environment through restrictions on the use of certain hazardous substances. Specifically, these materials are lead, mercury, cadmium, hexavalent chromium, polybrominated biphenyls and certain polybrominated diphenyl ethers. RoHS became law in the UK in August 2005 and the proscription of the identified hazardous materials applied from July 2006. The RoHS Directive has had, and continues to have, a significant impact on manufacturers, sellers, distributors and recyclers of electrical and electronic equipment. Producers need to ensure that the products they put on the European market do not contain the proscribed materials and that they comply with the requirements of the Directive. If a producer is found to have placed products that contain these proscribed materials on the European market they may be forced to withdraw them. The RoHS Directive covers all of the products categories described in the WEEE Directive, except for the medical and monitoring and control categories. Because it is not possible to eliminate every single atom of a substance, the RoHS Directive states that a material must not be present above a specified percentage weight in what is known as an homogenous material. This figure is set at 0.1% by weight for each of the proscribed materials, except cadmium for which the level is ten times lower at 0.01%.

Although the RoHS Directive only applies to products put on the market in European member states, it has encouraged the adoption of related legislation around the world. Perhaps the next most well known piece of this type of legislation is the so-called 'China RoHS', which proscribes the use of the same list of materials as the European RoHS Directive but which implements the requirements in a completely different way. More recently, Norway has announced that it is considering implementing its own version of RoHS which has been given the nickname 'super-RoHS' because it includes 18 distinct chemicals rather than just the 6 covered by the European RoHS Directive. This proposed Norwegian legislation is actually more correctly referred to as the 'Prohibition of Certain Hazardous Substances in Consumer Products', and is intended to be an additional chapter of the Norwegian Products Legislation. The Prohibition is directed at all products intended for consumers or reasonably expected to be used by consumers. So, although electronic and electrical equipment is included, it actually has a much broader scope.

3.4   *Other Examples of Legislation*

Although there are number of other pieces of Producer Responsibility legislation that may have some impact on the management of electronic waste, the two that are perhaps most likely to be of interest in the immediate future are the Energy-using Products Directive[8] and the REACH Chemical Regulations (more details on REACH can be obtained from the European Chemicals Agency in Helsinki[9]). The EuP Directive is a framework directive that harmonises requirements concerning the design of equipment. The eco-design component of the Directive requires manufacturers to consider the entire lifecycle of specific product groups and to assess the ecological profile of the equipment. This includes carrying out a lifecycle analysis of equipment which considers:

- raw materials
- acquisition
- manufacturing
- packaging, transport and distribution
- installation and maintenance
- use
- end of life

For each part of this process, manufacturers will be required to assess the consumption of materials and energy, emissions to air and water, pollution, expected waste and recycling/re-use. Thus, the EuP Directive (EuP) encourages the electronics industry to adopt a more holistic approach to the way it manufactures its products, with emphasis being placed on all aspects of a product's lifecycle from eco-design to end of life. The encouragement of eco-design principles will lead to the integration of environmental considerations during the design phase of a product, *e.g.* the best way to improve its environmental performance and to achieve more sustainable product development. Manufacturers and consumers should be able to benefit from better designed, more efficient products both economically and through the better use of finite resources. The European Parliament and the Council adopted a final text for the EuP Directive 2005/32/EC in July 2005. Actual measures are being decided on a product-by-product basis under the supervision of a designated panel of EU member state experts as part of the so-called 'fast-track comitology procedure'. Priority products include heating, electric motors, lighting and domestic appliances. Ultimately, this framework directive will cover all products consuming energy, apart from motor vehicles, and it is thought that these could account for 40% of the carbon dioxide emissions responsible for global warming, which are to be reduced under the Kyoto Protocol. New materials and processes will undoubtedly play an increasingly important role in helping to achieve legislative compliance.

The REACH (**R**egistration, **E**valuation, **A**uthorisation and restriction of **Ch**emicals) Regulations came into force on 1 June 2007 and they represent the new European system for regulating chemical safety. They will, potentially, have significant impacts not just on chemical producers but all the way down industry supply chains to end users. REACH will affect most businesses, including many that would normally not consider themselves involved with chemicals. There are expected to be ramifications for electrical and electronic products and there may well be a proscription applied to a number of materials in addition to those cited in the RoHS Directive. Although the list of these substances is not expected to be published until 2009, there are indications that certain flame retardants and plasticisers found in electrical and electronic products may be proscribed. The proscription of such materials could thus have an impact on the ability to recycle materials from products placed on the market before the REACH Regulations came into force.

For many large multinational companies one of the key challenges will be the integration of cost-efficient materials compliance strategies across multi-tiered global supply chains. With global supply chains often emanating from the Far East, there will clearly be an increasing need to minimise the level of supply chain confusion that could emerge as a result of the introduction of disparate legislation in different states, countries and regions of the world. There are already issues about the supply of RoHS-compliant components for certain regions and not others. Ideally, the industry will push for a greater degree of consensus with regard to environmental legislation, since one clear benefit would be a reduction in the need to manufacture region-specific products. It is important that new environmentally related legislation, wherever it is implemented, does not differ substantially in scope from other similar legislation in a different region. Achieving a degree of harmonisation through an international standardisation process will become increasingly important as more environmental legislation is implemented in coming years.

In attempting to achieve a harmonised approach to environmental regulation, particularly legislation governing materials restrictions, there will need to be greater engagement between industry and policy makers so that legislators understand the environmental trade-offs inherent in materials substitution. Additionally, industry must become more proactive in negotiating on how costs of sustainability best practice and resource efficiency can be absorbed into the economy without threats to inflation and employment.

Concurrently, there will also be a need to develop and standardise good scientific methodologies and testing procedures to demonstrate compliance as well as to assess the true environmental impacts of materials and potential trade-offs of alternatives. In the interim, or in the absence of such a harmonised approach, designers will require a comprehensive and standardised framework or 'quick-reference' system to enable them to keep abreast of developments in global materials restrictions and to assist them in determining the status of a given substance with respect to its environmental classifications in categories such as regulatory restrictions, EU risk phases and persistent organic pollutants, and to assist in materials selection and management.

4   Treatment Options for WEEE

Although there have been a number of developments in recent years on the treatment options for WEEE, there is still a need for additional new technologies that can further enhance the effectiveness of WEEE treatment in a more automated way that requires less manual intervention. Conventional treatment routes may typically involve some degree of manual separation followed by a variety of more automated comminution and separation techniques such as are described in more detail in later chapters. In order to help facilitate the more efficient treatment and recycling of electronic waste, the WEEE Directive defines ten categories of waste. However, during the run-up to the implementation of the WEEE Directive, there was a lack of definition around the specific details of the treatment requirements of WEEE. For example, there was uncertainty about exactly at what stage of the recycling process printed circuit boards (PCBs) and Liquid Crystal Displays (LCDs) would need to be removed from the waste stream. Similarly, there was also no definition of the exact number of different types of WEEE collection skips that would be available at Civic Amenity sites. The WEEE Directive defines ten individual categories of WEEE, yet the actual number of segregated streams that are in reality collected means that different types of products are not separated as efficiently as would ideally be needed for subsequent optimised treatment. This reduced number of segregated categories was proposed as a response to the need to limit the number of skips actually physically required at Civic Amenity sites because of space limitations.

It was, for example, concluded that it was not practical to have a skip for each of the ten different categories of WEEE, and simplification of the ten WEEE categories into five was recommended, with these covering:

- **Refrigeration equipment** – that requires specialist treatment under the Ozone Depleting Substances regulations
- **Other large household appliances** – that have a metal-rich content and can be easily reprocessed together
- **Equipment containing CRTs** – due to health and safety concerns relating to broken monitor glass this grouping must be handled separately
- **Linear and compact fluorescent tubes** – to prevent contamination and to enable easier recycling
- **All other WEEE** – those where there are no known technical reasons or EH&S concerns which prevent this mixed grouping of WEEE from being reprocessed together

Segregation of WEEE into specific, well-defined streams at the collection stage is clearly an effective approach technically for facilitating subsequent efficient recycling and re-use. In terms of the mixed grouping of 'all other WEEE', as one of the five categories, there appear to be no technical or Environmental Health & Safety reasons that prevent reprocessing together. However, the generation of highly mixed waste streams does not encourage re-use of components and recycling of added-value materials. Segregation of smaller electronic products would make recycling much easier, but little is still known of the technical solutions that may be available in the future and the economic balance is also far from established. Technologies that are capable of separating diverse and complex streams will therefore be crucial for maximising the recovery of resources from WEEE.

5 Material Composition of WEEE

One of the key factors that determines the choice of the most appropriate technology for recycling is the material composition of WEEE. There are not only clearly significant differences in the types of equipment that fall into each of the ten WEEE categories but even within individual types of products. The rapidly accelerating transition from CRT-based televisions and monitors to those employing LCDs is a very apposite example. The typical material contents of CRT, LCD and plasma televisions is shown in Table 2.

Also, even in terms of the materials that are common to many electrical and electronic devices, there are also changes being driven by legislation such as the RoHS Directive. The best known example is the transition from lead-based to lead-free solder, which has already occurred and which has been mandated for many products since July 2006. Whilst for some long-lifetime products, such as televisions, this will not have an impact for many years, the entry into the waste stream of short-lifetime products such as mobile phones will mean that the metals make-up of printed circuit board (PCB)-related scrap will start to vary and that there is likely to be a wider range of metals encountered than the tin, lead and copper typically associated with traditional circuit boards. Similarly, the proscription of cadmium, mercury and hexavalent chromium, as well as certain brominated flame retardants will also lead to compositional changes that will herald the introduction into the waste stream of a wider range of materials. This in turn will have ramifications for any recycling technologies that are developed to address individual waste streams for each of the WEEE Directive's ten categories. Conversely, it will also be increasingly important to segregate RoHS-compliant products from those that are non-compliant since, where recovery and recycling of materials is being undertaken for re-use in new electrical and electronic products, the presence of proscribed materials will make the treatment and recovery processes more complex and costly, in order to avoid potential contamination issues.

Table 2 Waste material content of CRT, LCD and plasma televisions.

	Material content/kg				
	Glass	*Metal*	*Plastic*	*Silicon*	*Total*
CRT	37	4.2	8	4.4	**53.6**
LCD	3.6	8.4	15	9.6	**36.6**
Plasma	14.8	12.4	10.9	8.6	**46.7**

Most types of EEE contain varying quantities and types of plastics and it has been understood for some time that there is a need to minimise the number of types used in electrical and electronic products in order to facilitate more effective recycling. The situation can be further complicated by the fact that there are compatibility issues, not only between individual classes of polymers, but also between the many different products that are produced for each class. The plastics that are commonly encountered in EEE are listed below:

- Acrylonitrile Butadiene Styrene (ABS)
- Polycarbonate (PC)
- PC/ABS blends
- High-impact polystyrene (HIPS)
- Polyphenylene oxide blends (PPO)

However, it should be noted that it is quite common to find many more types of materials used in specialist applications and electronics containing circuit boards will additionally contain a range of thermoset materials such as flame-retarded glass-reinforced epoxies. The ability to find uses for recycled plastics largely depends on the type of polymer, the cost compared to virgin material and the work required to produce a new material with the required purity and quality. For example, the need to separate the materials and to remove potential contaminants such as labels, screws and fixings can significantly increase the cost of recycled materials. There is also the need to consider the implications of recycling plastics containing brominated flame retardants. These are becoming increasingly unpopular because of their persistence in the environment and because of potential health and safety issues. Interestingly, in Japan, recycled plastics containing such flame retardants have found use as conduits for cables running along the side of railway lines.

The large range of plastic materials available and the incompatibilities that exist between them highlight the need for designers to reduce the number of types of material that are used in new products. In some instances, such as in laptop computer cases, manufacturers have sometimes used alternative materials including aluminium in order to assist with ease of recycling at end of life.

Available data on the specific material composition of EEE is both limited and disparate in nature. Information based on Japanese experiences detailed in a DTI Global Watch Mission report[10] shows material compositions for the four types of EEE covered by the Home Appliance Recycling Law (HARL); see Table 3.

Information has also been reported on the material composition of WEEE collected in Ireland during 2000; see Table 4.

One major difficulty from a materials recycling perspective is that it is still currently uncertain how end-of-life electronics will actually be segregated during the journey from the end user to the recycler. Clearly, there needs to be some definition of how products will be segregated, but best practice varies and is still evolving. Nevertheless, for the successful implementation of suitable, category-specific, focused recycling technologies, it will be important for the

Table 3 Materials composition (weight %) of the four products covered by HARL.

	Television	Washing machine	Air conditioner	Refrigerator
Glass	57	–	–	–
Plastic	23	36	11	40
Iron	10	53	55	50
Copper	3	4	17	4
Aluminium	2	3	7	3
Other	5	4	10	3

Table 4 Material arisings from WEEE in Ireland (2000).

		Arisings (tonnes)	
Material type	*Composition (weight %)*	*Upper*	*Lower*
Iron and steel	47.9	35249	15994
Aluminium	4.7	3459	1569
Copper	7.0	5151	2337
Other metals (non-ferrous)	1.0	736	334
Metals total	60.6	44595	20235
Flame-retarded plastic	5.3	3900	1770
Non-flame-retarded plastic	15.3	11259	5109
Plastics total	20.6	15159	6879
Glass	5.4	3974	1803
Rubber	0.9	662	301
Wood and plywood	2.6	1913	868
Concrete and ceramics	2.0	1472	668
Printed circuit boards	3.1	2281	1035
Other	4.6	3385	1536
TOTAL	**100**	**73589**	**35615**

individual waste streams to be characterised, both in terms of their product contents and their individual materials make-ups. Each recycling technology or process will have an optimum efficiency in terms of the raw material supply to be processed. Feedback about the material make-up required to achieve maximum efficiency would thus better enable those collecting and aggregating specific groups of products to control the composition in order to achieve enhanced efficiencies.

6 Socio-economic Factors

Socio-economic factors have a major influence on the viability of the overall approaches that can be adopted for treatment of WEEE. They also have a significant impact on the materials recovery processes that can be employed, as well as the specific techniques that are likely to be successfully implemented in the future. At present, and especially in the UK, there appears to be something of a disconnect between the desire for society to behave in a more sustainable manner and the introduction of new and more efficient WEEE recycling and recovery processes that can provide high-value recyclates suitable for real applications. For example, there is currently virtually no serious use of recycled polymers from end-of-life electronics, although companies such as Axion Recycling Ltd, who have recently commissioned a plant in Salford, have developed processes which are beginning to make it possible to provide high-quality recyclate. Conversely, in Japan, there is a more integrated approach to electronics recycling, which involves the major electronics companies as key players. This means that there are internal opportunities within vertical supply chains for the use of such recyclates. Technology exists for producing higher quality polymers but it is compromised by collection mechanisms that often lead to the basic granulated material being contaminated, not just with other plastics but also with metals, paper and dirt, *etc*. Part of this is due to the way business operates in Europe, with there being little connection between the manufacturers and the recyclers. This is the result of a combination of historical and cultural factors since, in Japan, the major electronics manufacturers actually own the recycling facilities and they are keen to source high-quality recyclate for re-use in their own products. Effectively, and unlike in Japan, internal supply chains don't exist in Europe and particularly in the UK. One key requirement, therefore, for WEEE recycling facilities in the UK will be to ensure that supply chains both before and after recycling are put in place in order to enable the processes to be optimised in terms of recovery, recycling and re-use of materials.

In order for industry to adopt the new recycling technologies that will inevitably be needed as legislation drives the recycling of more and more electronics waste, there will also need to be significant changes in the economics of recycling. The electronics industry is a truly global business that also extends to end-of-life recycling. This has been evidenced in recent years by the exports of WEEE from Europe, which have both caused controversy and embarrassed some well-known manufacturers of electronics. Other global factors also clearly have a major impact on what level of recycling activity takes place. For example, the values of many of the metals found in WEEE have increased significantly in recently years, largely in response to increasing demand from expanding economies such as China. The price of copper has tripled since the turn of the century and this offers an opportunity for newer technologies to be introduced that can enable more recycling to be carried out in the UK and which will, ideally, also generate greater volumes of high-quality recyclate. Whether these increases in the value of basic materials will be sustained will be one important factor influencing decisions about the expansion of, and investment in, metals recycling, which is currently a key part of the overall WEEE recycling process. Similar important considerations also apply to oil and energy prices, since these have an impact on the manufacture of raw materials used in electronics. They also influence the cost of transport for moving WEEE to the recyclers and indeed the cost of operating the recycling facilities themselves. Ideally, of course, WEEE should be transported as little as possible during its journey from being discarded, through the recycling and treatment process to the next user of the recovered material. From a sustainability perspective, a WEEE recycling facility should handle as much locally sourced WEEE as possible and supply the recyclate to local end users. This may not always be possible, particularly as there is a lack of integration in the process as mentioned above and because there may be limited local opportunities for the recyclate to be used. However, all of these potential recycling routes do need to be considered in a more holistic manner using appropriate lifecycle approaches. For example, in some cases the shipping of end-of-life electronics to the Far East for recycling and materials recovery might ultimately make sense if that is where the materials can be re-used to make new products.

The socio-economic factors that influence the need and ability to introduce new technologies into electronics recycling are also many and varied. They are also inextricably linked and thus there is a complex set of interactions and variables that impact what happens to WEEE. The need to achieve enhanced levels of re-use, recycling and recovery from UK WEEE is clearly demonstrated by comparisons with what is achieved in countries such as Sweden and Japan. The WEEE Directive is already requiring specific targets to be met for each category of WEEE and it is likely that these will become increasingly stringent in the future, with legislation such as the Energy-using Products Directive also likely to have an impact. This in itself will do much to encourage the adoption of enhanced and new recycling technologies. From a socio-economic perspective there is much more that could be done in the UK to help facilitate the introduction of these technologies and thus the improved processing of materials from WEEE. Ultimately, the recycling of materials is essentially driven by economics since, without the overall financial equations balancing, recycling would not be viable.

7 Logistics of WEEE

The introduction of Producer Responsibility legislation and the implementation of the 'Producer Pays' principle have, in recent years, led to companies funding the take-back and recovery of their products at end of life. The overall objective of these legislative measures is to reduce the quantities of discarded products being sent to landfill by setting mandatory recycling and recovery targets, the costs of which must be borne by the producers. The embodiment of this new approach to waste management legislation within the European Union has been aimed at packaging waste, end-of-life vehicles and, more recently, electrical and electronic products *via* the introduction of the Waste Electrical and Electronic Equipment (WEEE) Directive, which has been briefly described earlier. One of the critical issues in achieving these take-back requirements is the need to develop new, cost-effective and efficient logistical systems to enable the transport of WEEE from collection points to the recycling facilities.

The WEEE Directive identifies two types of WEEE waste streams: products discarded by private households and business-to-business (B2B) WEEE, as well as establishing a 4 kg collection rate *per capita* of WEEE to be achieved from private households. It is also encouraging that some electrical and electronic product retailers, such as Comet, have introduced take-back schemes that enable end-of-life products to be returned directly to their stores for subsequent shipment to recyclers. In terms of European implementation, a complication arises from the fact that the Directive is not a Single Market Directive and as such it has enabled individual Member State countries to establish the number, capacity, organisation and management of the collection points from which the producers will bear the costs of financing the separately collected WEEE. The freedom for European Member States to define the general requirements of their respective implementing legislation has resulted in different WEEE collection schemes being established across Europe. This has undoubtedly added an increased level of unnecessary complexity and cost to the producers and is in marked contrast to the approach adopted in Japan.

An overview of the different WEEE collection schemes across Europe can be found in the various issues of the Perchard's Report[11] that have been published in the last few years. During the implementation phase of the WEEE Directive consultation groups comprising producers, trade associations, compliance schemes and local authorities met with Government representatives in order to resolve how the allocation and transport of WEEE to recyclers could be best implemented in the UK. Consequently, and since the implementation of the Directive, the pathways for the collection and transportation of WEEE from collection points to treatment facilities have developed and evolved in the UK. However, waste management, transport and producer representatives have made it clear that the UK delay in transposing the WEEE Directive into legislation caused major issues in planning logistics, as well as making the appropriate investment in data collection and capital equipment, *etc.*, very difficult.

One of the more clearly understood decisions relating to WEEE allocation and collection was that Civic Amenity (CA) sites were to be defined as collection points, with the subsequent collection and recycling of WEEE falling to the producers, since many householders deposit large bulky household appliances at CA sites. Studies undertaken by Network Recycling[12] indicated that there was considerable capacity to expand WEEE collections at CA sites on the basis of available space alone and that such an approach would make a significant contribution in assisting in the implementation of the Directive. Network Recycling also estimated that total CA WEEE arisings were as high as 6.7 kg of WEEE/inhabitant/year. The current infrastructure for WEEE separation at CA sites is dominated by large household appliances with a high metal content and more than three quarters of the WEEE that is recycled comes from such appliances (washing machines, fridges, freezers, tumble dryers, dishwashers, *etc.*). This high recycling rate may be explained in part by the high value of the metals recovered compared to the much lower reprocessing costs which are further lowered because the facilities employed to shred end-of-life white goods are sometimes also used for recycling end-of-life vehicles. Other appliances that are separated at CA sites include refrigeration equipment due to the specialist treatment prescribed under the Ozone Depleting Substances Regulations. It should also be noted that there are also some small-scale programmes for separating small appliances to support re-use schemes.

In addition to requiring the set-up of efficient collection schemes that meet the WEEE Directive's targets, the Directive also requires that the collection and transport of WEEE should be implemented in such a way that supports the re-use and recycling of components or appliances that are capable of being re-used or recycled (Article 5.4 of the WEEE Directive). However, the reality of achieving cost-effective WEEE transportation dictates that as much product as possible is loaded onto a vehicle. This frequently can involve the use of heavy plant such as JCBs which can cause significant damage to the equipment. In contrast, one of the clear messages that came out of a report of the 2005 DTI 'Global Watch' mission to evaluate WEEE processing technologies in Japan was the difference in the Japanese approach to the handling and transport of WEEE. The Japanese approach differed in so much as that WEEE was regarded as a valuable source of raw materials, as opposed to being viewed as waste. The WEEE was streamed by item type (four categories of product) at the collection point and transported in specialist cages on specially designed hydraulic vehicles to minimise damage during transit. This method gave rise to higher than expected recycling rates and enabled higher yields of better quality materials. Whilst the requirements of the WEEE Directive are far more extensive than those required by the Japanese Home Appliance Recycling Law, more consideration will need to be given in Europe to sorting and handling issues if the opportunities for recycling and re-use are to be further improved.

Within the UK there appear to have been only a few studies made of the composition and amounts of WEEE generated. One example is work carried out by Axion Recycling for Enhance (the Enterprise support service from London Remade and London CRN) to characterise WEEE flows within the London area, inside the boundary defined by the M25 orbital motorway.[13] The study focused on the two primary sources of WEEE, namely that arising from the domestic sector, *i.e.* 'household WEEE' and that arising from business and commercial premises 'B2B WEEE'. The report by Axion Recycling noted that the UK is not unique in this lack of WEEE data. The Irish Republic (which implemented the WEEE Directive in August 2005) has also reported a scarcity of reliable pre-implementation WEEE data. It was also reported that there was an absence of unified reporting and calculation methodologies throughout the European Community. A similar lack of pre-implementation data was found in the Netherlands and Sweden. This suggests that the recording and reporting of WEEE volumes may only occur after implementation of local legislation and thus it may take some time for more reliable and accurate data to become available for the UK.

8   WEEE – the International Perspective

8.1   *European Perspective*

The other countries within the European Union all operate WEEE collection and treatment schemes with those such as Belgium, Denmark, Netherlands, Sweden and non-EU Norway and Switzerland being particularly noteworthy. The schemes are operated and owned by not-for-profit organisations involved in the manufacture and distribution of electrical and electronic goods. Denmark is the exception, where the WEEE collection scheme is operated by municipal and regional authority collectives. Austria and Ireland initiated collection schemes in 2005 and other EU member states have over the last few years transposed the WEEE and RoHS Directives into national law and registered collection schemes. Germany's collection scheme started in the first quarter of 2006. Some of the countries, such as the Netherlands, operate multiple schemes where different schemes collect different categories of WEEE. Table 5 below shows examples of the schemes operating in specific countries.

There are typically three main types of collection routes for end-of-life EEE and these are:

- Municipal collection sites
- Retailer take-back
- Producer take-back

All the schemes outsource the majority of their transport and recycling activities to commercial suppliers, usually on the basis of competitive tender contracts. Sweden collects the largest annual amount *per capita* of 14 kg, followed by Norway and Switzerland which achieve in excess of 8 kg. Belgium, Denmark and the Netherlands collect 4 to 5 kg. Before the WEEE Directive came into force, none of the countries collected all product categories covered under the Directive and hence they were required to expand the product collection areas in which their respective schemes currently operated. An example of the composition by weight of WEEE collected, in this case in Ireland, is shown in Table 6.

Table 5   Example European recycling schemes for WEEE.

Country	Scheme
Belgium	Recupel
Denmark	Target Municipal Tax
Netherlands	ICT Milieu
	NVMP
Norway	El Retur
Sweden	El Kretsen
Switzerland	SWICO
	S.E.N.S.

Table 6 Composition of WEEE by weight collected in Ireland.

Product type	Composition by weight/%
Large household appliances	79.9
Consumer equipment	12.5
IT and telecoms	3.6
Small household appliances	3.5
Electrical and electronic tools	0.4
Lighting equipment	0.2
Toys, leisure and sports equipment	0.002

Table 7 Product coverage based on WEEE categories.

Category	Belgium Recupel	Netherlands ICT/NVMP	Norway El Retur	Sweden El Kretsen	Switzerland SWICO/ S.E.N.S.
Large household	Yes	Yes	Yes	Yes	Yes
Small household	Yes	Yes	Yes	Yes	Yes
IT and telecomms	Yes	Yes	Yes	Yes	Yes
Consumer equipment	Yes	Yes	Yes	Yes	Yes
Lighting equipment	No	No	No	Yes	No
Electrical and electronic tools	Yes	Yes	No	Yes	Yes
Toys, leisure and sports	No	Yes	Yes	Yes	No
Medical devices	No	No	Yes	Yes	No
Monitoring and control	No	No	No	Yes	No
Automatic dispensers	No	No	No	No	No

Table 7 highlights the product coverage on the basis of WEEE Directive categories for various countries' compliance schemes.

Prior to the introduction of the WEEE Directive, concerns were voiced about the cost of compliance. However, with the benefit of experience following implementation, more recent indications are that recycling costs in relation to the end-product price for consumers were lower than initially anticipated. Significantly, in countries where there is a more competitive electronics recycling environment, the take-back and recycling costs of electronic equipment are lower. Costs are generally higher in countries where there is no competition and only one recycling provider for the industry to work with. For example, Austria, Germany and Spain, which have relatively new and highly dynamic take-back and recycling systems with strong market competition, have been able to achieve costs of as low as a few Euro cents per product. Calculations for notebook computers indicated a cost of around 7 € cents in Germany, 20 €

cents in Spain and 39 € cents in Austria. Conversely, in Belgium, Switzerland and Ireland, where competition was more limited, the costs were higher. This should mean that, with the relatively large number of producer responsibility schemes in the UK, prices should also be highly competitive.

It can thus be seen that, despite the implementation of the WEEE Directive across all member states of the EU, there are clearly differences in the practical approaches actually implemented in different countries. It will be interesting to see how these differences evolve in the future.

8.2   Japan

Although Japan is well known for its proactive approach to electronics recycling, it has taken a markedly different approach to the one employed in Europe *via* the WEEE Directive. It was the implementation of the Home Appliance Recycling Law (HARL) in April 2001 that really proved to be the key driver for the significant level of recycling activity established in recent years. The HARL covers four major types of home appliances: televisions, refrigerators, washing machines and air conditioners. It is thus both different and simpler than the individual national schemes that are being proposed and implemented within the UK and other European countries in response to the WEEE Directive. The HARL adopts a completely different approach to the implementation of end-of-life recycling, where the consumer pays a fee when requiring disposal of an appliance falling within one of the four categories covered. Specifically the law requires that:

- consumers pay a recycling fee when disposing of home appliances
- retailers take back discarded appliances and pass them on to manufacturers
- manufacturers recycle the recovered discarded appliances

Interestingly, the process works well with large quantities of appliances being recycled and a high proportion of the materials recyclate finding new uses. This, in turn, indicates that individual consumers are indeed willing to pay the required recycling fee when an appliance is sent for recycling. It is difficult to believe that this type of approach would work well in the UK, since there is already widespread dumping of WEEE away from the proscribed recycling centres and Civic Amenity sites, as can be seen in Figure 3.

The significant cultural differences between the UK and Japan suggest that the implementation of such a scheme based on a consumer fee in the UK would simply lead to an increase in the illegal dumping of even more quantities of WEEE.

In order to handle and process the end-of-life appliances covered by the HARL, 46 home electric appliance recycling plants are operated throughout Japan and they employ approximately 2200 people. Of the four specified types of end-of-life electric appliances brought into these home appliance recycling plants from designated collection sites nationwide during 2003, there were 3.57

Figure 3 Abandoned electrical and electronic waste in Warwickshire.

million televisions, 2.68 million refrigerators, 2.68 million washing machines and 1.59 million air conditioners. This gave a total of 10.52 million units and a composition ratio of 35%, 25%, 25% and 15%, respectively.

One key element in the successful implementation of the Home Appliance Recycling Law is the use of a ticketing or voucher system that ensures the traceability of an end-of-life appliance on its journey from the consumer to the recycler. The ticketing system has been found to be a very useful way of monitoring and tracking appliances at end of life. The system is administered by an organisation known as the Home Appliance Recycling Law Ticket Centre, a public foundation that is owned by the Japanese Government. (The ticket centre is regulated by the HAPA, a trade association of appliance manufacturers, but it is ultimately under government control.)

As with the take-back requirements for WEEE in Norway, Sweden, the Netherlands and Belgium, the Home Appliance Recycling Law imposes an 'old for new' requirement on Japanese retailers. This means that every time a product is sold, one must be taken back from the consumer. This can be either a similar used product or another product that sold in the past. The law also permits manufacturers to contract with other organisations, such as the Association for Electric Home Appliances (AEHA), to provide collection services on their behalf. In rural areas without major appliance retailers, collection is provided by the local government or the AEHA. The Japanese Ministry of the Environment has estimated that 80% of recycled appliances are being collected through retail outlets. Following collection, retailers, local government and other designated organisations are obligated to transport the collected appliances to consolidation centres operated by two consortia of manufacturers. The HARL specifies that manufacturers have individual responsibility for their products and the Japanese industry has thus formed these two consortia to address this responsibility. The manufacturers in each consortium are also required to establish regional consolidation centres and to ensure the transport of collected products from these centres to recycling facilities. Each consortium operates approximately 190 consolidation centres, as well as a significant number of the recycling facilities.

The first consortium, known as Group A, has amongst its membership the following manufacturers: Electrolux, GE, Matsushita and Toshiba. The second, Group B, includes Daewoo, Hitachi, Sanyo, Sharp and Sony. Companies that sell only a limited number of products in the Japanese market are allowed to designate other organisations to fulfil their collection and recycling responsibilities on their behalf. As stated above, the recycling process is funded by the consumer who is required to pay a fee when an appliance is consigned for recycling. This is a completely different approach from the up-front European funding model, where the costs of recycling are included in the price of the product and are invisible to consumers or are shown in advance as recycling fees that are identified on the product receipt at the time of purchase (for example, as is popular in France).

In Japan consumers pay a collection fee, set by the retailer or other collection agent, when they take their used products for recycling. They also have to buy a recycling ticket that can be obtained from post offices and some retail stores; see Figure 4.

The basic process requires the use of a system employing five tickets per item to be recycled and these are used to track the progress of the appliance through the recycling system. Each 'ticket' consists of a book of five copies printed with an individual number and, in total, costs the equivalent of between £8 and £19, the exact amount varying with appliance type. The actual costs in US$ (depending on exchange rate) are reported to vary as follows, with the highest charges being made for orphan products:

- Air conditioners: $23 – $30
- Refrigerators: $30 – $38
- Televisions: $18 – $24
- Washing machines: $16 – $22

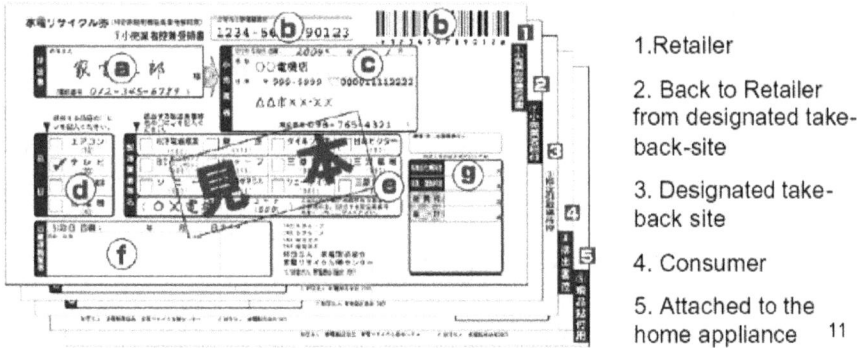

1. Retailer

2. Back to Retailer from designated take-back-site

3. Designated take-back site

4. Consumer

5. Attached to the home appliance   11

Figure 4   The tickets used in the Japanese recycling system under HARL.

The recycling fees collected are transferred to the manufacturers on a monthly basis. They are intended to cover the costs associated with operating the regional consolidation centres, transporting products to recycling facilities and the recycling operation itself. However, the fees are insufficient to cover all the costs of recycling and hence the manufacturers are responsible for funding the remaining costs.

The tickets contain details of the appliances, the name of the retailer and manufacturer and, from the information and reference number on the ticket, consumers are able to check the status of their appliance, *e.g.* to find out whether it has been returned to the manufacturer or sent on to a third-party recycler. Consumers thus have full traceability of their appliance *via* this relatively simple system. They are able to find the status of an electrical appliance submitted for recycling by accessing the web page of the AEHA and using a checking system that tracks the status of all collected electric appliances. (The AEHA is also responsible for orphan products, *i.e.* those that outlast the manufacturer, such as a TV set discarded 20 years after the date of sale.) The ticketing scheme places a number of obligations on people and organisations in order to ensure that the process works efficiently. For example, consumers are obliged both to cooperate in appropriately transferring used appliances to retailers in order to ensure recycling and to agree to pay the necessary fees for the transfer and recycling of those appliances. Similarly, retailers have an obligation to take back used home appliances, both when the appliances are those which retailers themselves previously sold to consumers and when retailers sell the same kind of home appliances to consumers. They also have an obligation to transfer them to the relevant manufacturers or importers. Manufacturers and importers are then obligated to receive appliances for recycling at designated take-back sites and to recycle them, according to the recycling standards set by the Japanese government.

The recycling of the four categories of appliances impacted by the HARL takes place with a high degree of efficiency. It is interesting to make comparisons between the process the Japanese have successfully implemented to treat WEEE and the more complex situation that prevails in the UK and the rest of Europe. One of the key differences between the European approach to WEEE recycling and the Japanese approach is that the Japanese model only addresses four specific types of WEEE. This is in marked contrast to Europe, where the WEEE Directive defines ten categories of products and seeks to embrace virtually all electrical and electronic waste. Although the Japanese have targeted a much narrower range of products than in Europe, their government estimates that the four product categories addressed by the law account for 80% by weight of all discarded electrical and electronic equipment. It is also interesting to note that there is no equivalent complimentary Restriction of Hazardous Substances Directive-type legislation in Japan, although it should be highlighted that the Japanese were among the first to adopt lead-free manufacturing for their electronics, way ahead of any legislation either in Japan or Europe.

9   Barriers to Recycling of WEEE

One of the main barriers to recycling of WEEE by manufacturers is the distribution of WEEE in relation to the location of the manufacturing plant. Many WEEE manufacturers are based a significant distance from their markets and also from the resulting waste. This makes it difficult and expensive for them to operate take-back facilities specifically for their appliances.

Retailers and distributors are in the best position to collect WEEE, as old appliances can be collected when delivering the new ones, or people can take small appliances to their local store when buying/collecting a new appliance. The barriers to such schemes are that the retailers would need to carry the cost for collection points and provide storage facilities, which, apart from space constraints, would also have security and health and safety implications.

The local authorities in Scotland are major collectors of WEEE. Special uplift schemes and Civic Amenity sites are used by householders for the disposal of large bulky items such as white goods and larger TVs. However, a major barrier to this type of WEEE recycling is public apathy and the unwillingness of householders to go to any effort to separate their waste. This would require a large capital investment for the recycling facilities such as drop-off points, and a campaign to educate the public so that they will use any facilities provided. The main problems experienced by the scrap metal dealers and material recyclers are that the appliances are made from mixed materials, which makes disassembly labour intensive and therefore costly. The resulting metal is often contaminated with plastic and other insulation materials, *etc.*, which leads to a poor-quality product. This means that there is not a ready market for such products and so it often does not make economic sense for these appliances to be processed.

10 The Recycling Hierarchy and Markets for Recyclate

In the past, any reference to the recycling of WEEE would typically have really only been about the recovery of metals. This has been especially true for certain types of electronics products which contain valuable quantities of precious metals such as gold and palladium, as well as others including copper, tin and silver. Even with larger, high-volume electrical products, *e.g.* washing machines, the term recycling has really meant recovery of steel and copper. However, in terms of both more sustainable approaches and recovered value, there is a clearly defined hierarchy that can be applied when treating WEEE. This may vary somewhat, depending on the specific product type, but in general terms the preferred order of approach is as follows:

- Refurbish and re-use
- Repurpose
- Recover and re-use functional modules
- Recover and re-use components
- Recover materials
- Produce raw material feedstocks
- Recover energy

Therefore, within the overall WEEE re-use, recovery and recycling hierarchy, the key opportunities are for the re-use of individual functioning units, modules and components. There are now many organisations that are actually refurbishing electronic products for re-use and this has been particularly evident with mobile telephones and personal computers. These organisations are often operated as social enterprises in order to offer low-cost access to products for those that would not otherwise be able to afford new products. This has enabled wider computer and internet access to people both in the UK and, for example, to school children in the poorer parts of Africa.

Where it is not possible to refurbish and re-use electronic products, another interesting and relatively new concept involves the repurposing of certain types of electronics. Although this approach is yet to be widely exploited, the increasing sophistication of certain types of electronics, such as mobile phones and portable gaming equipment, means that they are potentially suitable for use in other applications. In the case of mobile phones, the latest products contain very advanced processing and communications capabilities that could be harnessed for new applications yet they are often discarded after very short use lifetimes of typically between 12 and 18 months. Considering that there were over 2 billion mobile phone users globally in 2005, this represents a waste of large quantities of valuable components, raw materials, energy and electronics functionality. Lifecycle assessment (LCA) studies have found that one of the most significant impacts of a mobile phone on the environment is the energy used to manufacture the phone's components. For example, manufacturing just one 32 Mb RAM module requires 32 kg of water, 1.6 kg of fossil fuels, 700 g of gases and up to 72 g of different chemicals. The premature disposal of these high-performance devices is not sustainable and can also lead to potential environmental issues associated with traditional disposal routes.

Recently, work has been carried out to examine the feasibility of repurposing mobile phones into generic smart processing units for re-use by manufacturers of new electronic equipment. By repurposing redundant units and converting them into general purpose modules, they can be used as core components in a wide range of applications such as those requiring smart monitoring, sensing and telemetry capabilities. Applications could range from burglar alarms and monitoring, to advanced mobile medical diagnostics for use in remote locations. While this approach is only applicable to a number of sophisticated, high-volume electronic products, it does offer interesting possibilities that avoid the conventional disposal and recycling routes and it is clearly worthy of further investigation.

Where refurbishment and repurposing are not appropriate, it may be possible to recover certain functional electronic components before consigning the rest of an electronic assembly to more conventional recycling. This allows additional value to be recovered from, for example, the assembled printed circuit boards found in most products. In Austria, the organisation SAT developed an automated component disassembly methodology for the dismantling of components from scrap, redundant or malfunctioning PCB assemblies. SAT's technology essentially comprised an automated optical recognition scanning system that identified components suitable for recovery (*i.e.* with a certain value) and a laser desoldering unit coupled with a robotic vacuum arm for the removal of selected components. After removal and testing the recovered components could then be re-used.

The NEC Group in Japan has also carried out work to develop an automated disassembly process using a mechanical approach. Equipment was developed to remove components in a conveyorised mode *via* heating with infrared and shearing and, as a separate development having a higher throughput rate, *via* crushing with impacting rollers. While both of these approaches left the circuit board intact, the former method resulted in removal of both surface mount and leaded components without loss of integrity. NEC additionally extended the heat-impacting equipment to effect residual ($\sim$4%) solder removal *via* automatic belt sanding.

However, while the approaches briefly mentioned above are very interesting, they will, in all probability, be limited to a relatively small proportion of the total volume of electronics waste. For much of the large volume of electronic waste products these routes will not be viable as many types of WEEE do not really contain the type of items that are economic to recover and re-use. Typical examples here would be the wide range of low-cost consumer goods such as kettles, toasters, radios, CD players and related products. These have no major electronic components of any value and, as such, they are likely to be processed through routes that will not vary significantly from those that have been practised for many years. These essentially involve the use of a range of comminution and sorting techniques that can separate the metallic fraction from the plastic fraction. More sophisticated techniques will enable ferrous and non-ferrous metal fractions to be segregated, but ultimately the metals will find their way to a refiner who will separate the elements to produce high-purity recycled metals. The plastic fractions can also be separated to a certain degree into individual types if there is a will to do so, but in many cases these materials may also end up being consigned to furnacing as part of the metal refining process. A key question that needs to be addressed, therefore, is whether or not these materials can be better segregated and re-used in alternative and novel applications, rather than them simply being sent to a refiner.

There is clearly a significant opportunity for the re-use of recovered and recycled materials but the key to success is the way in which they are handled and treated at end of life but before recycling. Materials that can be kept clean and uncontaminated are much more valuable than those that have been mixed with other materials and which may be both contaminated and dirty. In Japan, end-of-life electrical and electronic goods are treated as a valuable source of raw materials and end-of-life products are supplied to dismantlers and recyclers in relatively good condition. This enables individual parts to be hand disassembled and sorted for recycling. There are many opportunities for polymer recycling if a good enough quality recyclate can be obtained. It can, for example, be recompounded and re-used directly or mixed with virgin materials to give a range of products with properties virtually identical to new material. One of the big challenges with plastics from end-of-life EEE is what to do with those materials that contain brominated flame retardants. The Japanese have used recycled flame retarded plastics from WEEE to produce cable trunking for housing signalling cables along the side of Japan's railways. These large modular mouldings use a significant amount of recyclate, offer an alternative to the traditional concrete products and provide a stable material that has a life expectancy of many years. There is also the additional benefit of having a flame-retarded material. Alternatively, techniques have been developed for removing brominated flame retardants from end-of-life plastics and for sorting polymer recyclate to remove bromine-containing materials. One relatively simple approach for determining the presence of brominated flame retardants in polymers uses a sliding spark device which vaporises a small amount of the polymer. The resulting changes in the colour of the spark can then be used to identify the presence of brominated flame retardants, halogens and heavy metals such as cadmium and lead. This type of hand-held unit can be a very useful tool for polymer recyclers. Alternatively, a more sophisticated approach has been studied by workers at Surrey University, where a range of spectroscopic techniques has been combined with multi-variate statistical analysis to enable the rapid characterisation of mixed polymer waste streams.

One recent example of the successful recycling of polymers from end-of-life electronic products comes from Axion Polymers, based in Salford, which has developed a new RoHS-compliant grade of high-impact polystyrene that is manufactured entirely from old TV and computer-monitor plastic. Axion's unique process removes the fraction that contains brominated flame retardants from discarded casings to produce a purer material called Axpoly PS02. An example of a product manufactured from this material is shown in Figure 5.

It is also worth noting that a company called MBA Polymers UK has recently announced plans to build a 60 000 tonne capacity plant that will recover plastics from upgraded shredder residue. Shredder residue is a complex plastics-rich mixture of materials resulting from the recycling of cars, electrical/electronic appliances and other metal-rich streams. It has been estimated that the UK generates around 3 million tonnes of plastics from cars, electronics, electrical appliances and other end-of-life products that are disposed of each year. The plastics from them have typically been landfilled or incinerated at high economic and environmental costs because the recovery processes are viewed as being too complicated or expensive to operate. It has been calculated that, for every tonne of virgin plastic replaced, between two and three tonnes of the greenhouse gas CO_2 is prevented from entering the atmosphere.

For the future, there is growing interest in the use of non-fossil-fuel derived plastics in electronics applications such as mobile phone cases. For example, polylactic acid (PLA), a bioplastic derived from corn, has been generating growing interest in recent years as a possible replacement for conventional petroleum-based plastics. Until recently, PLA had not been used in electronic devices as it had insufficient heat resistance and strength. However, NEC and UNITIKA jointly developed a PLA bioplastic reinforced with kenaf fibre.[14] (Kenaf is a biomass-based flexibiliser and reinforcing filler.) This new material has already been used for the entire casing of an NTT mobile phone which was launched in Japan during March 2006.

Another unusual and novel recycling application for materials from electronics applications was developed by the German PCB company Fuba

Figure 5   Kitchen cabinet foot moulded in a RoHS-compliant high-impact grade polystyrene made entirely from recycled TV and computer-monitor plastic. (Courtesy Keith Freeguard, Axion Recycling Ltd.)

GmbH. The Fuba Gittelde PCB manufacturing facility also housed an 'all dry' metal recovery processing plant which was fully automated and capable of separating various metals, including copper and gold, from end-of-life and scrap circuit boards. While the process was primarily concerned with the separation and recovery of metals for refining, the resulting polymer-rich fraction, which was essentially a metal-free epoxy glass-fibre composite powder, was also recycled and re-used. In this case, Fuba used the material to make chemical-resistant pallets that replaced wooden pallets in other parts of their facility where acids and other etchants were used.

One big problem area for electronics recyclers is associated with end-of-life CRT-based televisions and monitors. The changeover to LCD technologies has resulted in the demise of traditional CRT manufacturing and thus one of the main uses for the lead-bearing recycled glass has rapidly declined. There has been a number of attempts to find novel uses for CRT glass and one UK company, Nulife Glass Ltd, has developed a process that extracts molten lead metal to provide a valuable stream of lead-free glass and metal that can be re-used elsewhere; see Figure 6 below.

Figure 6   An end-of-life cathode ray tube, granulated CRT glass and a product made from this glass after the lead has been removed. (Courtesy NuLife Glass Ltd.)

It has also been shown that glass from end-of-life CRTs can be used in the ceramics industry as a raw material in glazes for both low- and high-temperature fired products. End-of-life CRT glass has also been used as a component in concrete mixes. This use encapsulates the lead-containing glass from the CRT by using it as a replacement for sand to make a concrete composite. Test results have shown that this method can prevent 99.99% of lead from leaching into the surroundings. When the glass is combined with biopolymers, such as guar gum, and a cross-linking agent, concrete composites can be produced that have compressive strengths of up to 6000 psi.

There have been various products commercialised that are made out of scrap or end-of-life circuit boards. Most notable among these are clocks, coasters, aluminium-framed briefcases with panels made of circuit boards and ring binders whose sides are also circuit boards. The British company Ecotopia offers a range of such products made out of recycled printed circuit boards, examples of which are shown in Figure 7.

Another small company was recently advertising lampshades by the French designer Luc Gensollen that were made from end-of-life and reject circuit boards; see examples in Figure 8.

However, while these more unusual applications for end-of-life electronics and their materials are undoubtedly interesting, they are unlikely to have any major impact on treating the total volume of WEEE being produced around the world.

In summary, the key to the further use of recycled materials lies with the need to have a more-informed approach to end-of-life product segregation which may have to begin as far back as the design stage. This not only means separating products and keeping them in good condition prior to dismantling and recycling but also designing products that are easier to treat at end of life, with fewer types of materials and fewer fasteners and fixings. Simply shredding mixed WEEE items in bulk, as has been the traditional, minimum-cost route pursued across Europe in the past, prevents opportunities being developed to maximise recycling rates and thus to promote additional recycling of valuable materials.

11   WEEE Health and Safety Implications

Electrical and electronic products contain a wide range of materials and some of these are known to present potential health and safety issues for workers involved in their treatment at end of life. Even though the RoHS Directive has proscribed the use of several materials deemed to be hazardous from July 2006, they will still appear in waste streams for many years to come as legacy products reach end of life and need to be treated by recyclers. From a health and safety point of view there are several ways in which end-of-life electronics can present a hazard both to the people employed by the recycling industry and, potentially, also to the environment. Examples can include the exposure of recyclers to hazardous materials by the use of inappropriate treatment

Figure 7 Products made from recycled PCBs. (Courtesy Ecotopia, www.ecotopia. co.uk.)

techniques and the failure to use the required personal protective equipment. Improper treatment and disposal can also lead to contamination of the environment.

For a number of years there has been a growing move to replace conventional cathode ray tube- (CRT)-based televisions and computer monitors with newer types employing liquid crystal and plasma-based display technology. It is estimated that the UK disposes of around 2 million CRT-based televisions each year plus an additional quantity of CRT-based monitors, generating in the region of 100 000 tonnes of CRT glass per year. Consequently, there are large numbers of CRTs which need treatment and yet which contain a number of hazardous materials. CRTs contain a range of hazardous materials including a

Figure 8 Table lamps made using recycled PCBs. (Courtesy Luc Gensollen, lucgensollen@wanadoo.fr.)

number of potential carcinogens such as lead, barium, phosphors and other heavy metals including cadmium and mercury. When treated properly in a controlled environment, these do not pose any serious health or environmental risks, but the breaking, separation and disposal of CRTs in an uncontrolled environment without the necessary safety precautions can present significant hazards for workers employed in this type of process, as well as leading to the release of toxic materials into the environment, including soil, air and water courses. A DTI-supported Global Watch Mission that visited numerous European recycling facilities reported a number of examples of poor CRT treatment and disposal practices, with some workers being seen to simply smash CRTs with a hammer. Conversely, Japanese recyclers are known to adopt a much more controlled and safer approach. There is a wide range of recycling methods used for treating end-of-life CRTs, from those that clearly lack the necessary controls and procedures to prevent exposure of workers and contamination of the environment to those that are much more benign. With the typical lifetime of a conventional CRT-based television being in the region of 14 years, large numbers of CRTs will be appearing for recycling for many years to come.

With the move to LCD and plasma-based displays, the types of materials found at end of life will be different from those encountered with conventional CRT-based displays. Whether the situation will be better or worse from a health and safety and environmental perspective will vary depending on a number of factors. For example, there is a move to larger area TVs and displays which may mean larger quantities of materials will have to be treated. However, there seems to be a general consensus that these newer types of displays will be less problematic than conventional CRT displays. They are, for example, unlikely to contain lead in anywhere near the quantities found in CRT glass and those put on the market after July 2006 will be lead-free. LCD displays will also contain fewer heavy metals and they have no need to employ the phosphors that are found in CRT displays. It should be noted, however, that many LCD displays use mercury-containing backlights which will need to be treated in a safe way. There are currently a number of approaches of treating mercury backlights but it is often difficult to remove them from the rear of a display without them breaking and exposing workers to both mercury and the phosphors used to generate the light in the tube. Some large-area LCD displays contain more than ten mercury backlights and thus designers need to address the problem of removal at end of life by providing designs that enable their easier removal. Interestingly, the liquid-crystal chemicals used in these displays have largely been shown to be of very low toxicity. Based on the results of tests to determine the eco-toxicology of liquid-crystal materials it has been suggested that no special requirements are needed for the disposal of LCD displays. Thus, if the issue of the mercury-containing backlight can be addressed by their replacement with newer technologies, LCDs may have less health and environmental impact than the CRTs they replace. It is also worth noting that, in some applications, mercury backlights are being replaced with alternative light sources and thus it is possible that mercury backlights may soon no longer be used.

Another area where there are potentially both health and environmental issues is in the treatment of end-of-life electronics containing brominated flame retardants (BFRs). BFRs undoubtedly serve a very useful purpose and are responsible for saving many lives that would otherwise have been lost in fires, but they have also attracted growing attention in recent years because of their persistence in the environment and their potentially negative impacts on human health.

There are various types of BFRs found in electronics applications, including those that are reacted into thermosetting polymer systems such as the epoxides used in circuit board laminates and the non-reactive polybrominated diphenyl ethers (PBDEs) that are used in thermoplastic polymers. The levels of PBDEs in humans have been found to be doubling every 2 to 5 years, and are much higher in North America than on other continents. Despite the fact that certain PBDEs have been proscribed by the RoHS Directive, it is likely that BFRs will continue to be found in electrical and electronic products for some years to come. While the reactive BFRs, such as those based on tetrabromobisphenol A, have relatively low volatility, PBDEs are known to migrate out of polymers, especially when they are heated. Consequently, any recycling processes that use heat may lead to emissions of PBDEs and thus there is the possibility that both recycling workers and the environment may be impacted by these materials. For example, research in Sweden found that levels of deca-BDE in the blood of workers recycling electronic equipment were much higher than those in people who were simply using computers. Interestingly, there is also evidence that restrictions on the use of brominated flame retardants can increase the risk of environmental damage occurring due to fires in electrical products that are no longer, or less effectively, flame retarded.

As there is a growing interest in recycling and re-using polymers from end-of-life electronics, the issue of brominated flame retardants is one that needs to be addressed. For example, there is a need for careful controls to be put in place for workers involved in the recompounding and recycling of polymers containing these materials, since processes such as extrusion expose the polymer to high temperatures that can lead to the emission of BFRs and their degradation products. Questions will also have to be answered as to the eventual markets for polymer recyclate containing BFRs since they are likely to be increasingly avoided by traditional European users.

In recognition of the growing concerns around the use of brominated flame retardants, there have been moves by the industry to develop a coordinated approach to best practice. A good example of this approach is given by the Voluntary Emissions Control Action Programme (VECAP),[15] which was established by the brominated-flame-retardant industry. VECAP was set up to manage, monitor and minimise industrial emissions of brominated flame retardants into the environment through partnership with Small and Medium-sized Enterprises (SMEs). It effectively seeks to introduce best practice ahead of what will be required *via* the REACH regulations. Up to now the focus has been on deca-BDE with manufacturers and users working together to establish and share best practices on handling in order to minimise emissions to the environment. There is a VECAP Code of Good Practice and it seeks to achieve level reductions throughout the manufacturing process *via* a culture of continuous improvement. It is claimed that since the introduction of this programme, levels of deca-BDE in the environment have fallen.

While one of the key objectives of the WEEE and RoHS Directives is to reduce the environmental and health impacts of end-of-life electronics, one area where there are still significant problems is with the export of WEEE to developing countries such as China and Africa. Although this is, in theory, proscribed, such activity is still occurring and considerable quantities of scrap electronics are being treated in these countries in less than ideal conditions that are exposing both the recyclers and the environment to hazardous materials. For example, in many parts of Asia electronics recycling remains largely unregulated and is also poorly understood with regard to its impacts on the environment and on the health of recycling workers and surrounding communities. In many recycling operations the end-of-life electronic devices are subjected to mechanical comminution and shredding which can generate large quantities of particulates and dust that can easily disperse into the local environment.

Work has also been carried out to determine the levels of metals and other materials in dusts collected in the area around various Chinese and Indian electronics dismantling facilities and in many cases the levels of metals such as copper, lead and tin were very much higher than the expected background levels. Preliminary results from a study carried out in Beilin (China), in which dust samples were collected from the houses of two solder-recovery workers and from one household having no connection with the industry, indicated the potential for the home environment to become contaminated with chemicals from the workplace (*e.g.* as a result of contamination of work clothing). Although the houses were remote from the solder-recovery works themselves, levels of copper, lead, tin, antimony and, to a lesser degree, cadmium were higher in the dusts from the two solder-workers' houses than in the single control house sampled. These results illustrate the need for careful control procedures to be put in place in order to prevent the possibility of exposure to harmful materials from WEEE in the homes of recycling workers.

12 Future Factors That May Influence Electronic Waste Management

It is fairly certain that over the next ten years Europe and the rest of the world will have to be far more aware of their material and energy resource consumption. This will have a number of implications, both in terms of 'top-down' processes driven by legislation and compliance criteria, as well as by 'bottom-up' processes driven both by Corporate Social Responsibility and Sustainability requirements. One essential prerequisite will be the need to have better data tracking systems for all major flows of resources, *i.e.* energy and materials, both through supply chains and through society as a whole. The current absence of this type of key performance indicator information has been a significant contributory factor to why so little is presently known about waste in society. The effects of Producer Responsibility legislation will also continue to have a growing and noticeable impact, not only in Europe but increasingly throughout the rest of the world. Examples of this impact include moving the financial responsibility for waste management back up supply chains with the cost for recycling and recovery incorporated into the selling price of products, waste minimisation measures introduced at the design stage and changes in the material composition of EEE waste as producers utilise fewer and simpler materials in order to reduce end-of-life reprocessing costs. Additionally, there will need to be a greater convergence between the electronics industry and the waste sector. Strategic partnerships with reprocessors will need to develop and strengthen. As this legislation continues to be implemented and reinforced, the electronics industry will have to keep pace with the changes required. Some of the issues with which the industry will need to come to terms are:

- restrictions and limitations on the types of substances and materials that can be used in electronics, *e.g.* brominated flame retardants and heavy metals
- stricter end-of-life recovery, re-use and recycling requirements, *e.g.* the WEEE Directive
- enhancements in the energy efficiency of products in terms of both the manufacturing and use phases, aided by new materials technology and driven by legislation such as the EuP Directive
- adoption of more holistic approaches embracing all aspects of the product lifecycle from design through to end of life
- the adoption of more sustainable business practices
- a move away from the commoditisation of electronic products towards product service system-type approaches

Although it seems that the world is likely to witness the continuing evolution of existing silicon-based semiconductor technology for a number of years to come, there will undoubtedly also be major changes in technology that will have a significant influence on materials, processes and end-of-life considerations. We have already witnessed the demise of the conventional CRT-based display in most TV and computing applications and the new LCD displays that have replaced them utilise markedly different types of materials and in quantities that each bring their own end-of-life challenges and opportunities. If anything, this type of evolution will continue and the emergence of polymer and printed electronics will require the use of many types of new materials. There will also be continuing integration of electronics functionality at all levels and electronics is becoming ever more ubiquitous, bringing new applications and products in areas that were previously unimaginable. The age of 'electronics everywhere' is approaching and with it will come a demand for many new types of materials, each of which may require specialised and optimised waste management strategies.

13 Summary and Conclusions

This chapter has sought to introduce the subject of electronic waste and its management in the context of our global need to behave in a more sustainable manner. It has also attempted to set the scene for the more detailed coverage of specific topics in following chapters. It is clear that while the world population's growing access to ever-increasing quantities and types of electronic and electrical equipment has a number of major benefits, there are also serious issues that need to be addressed, not just at end of life but throughout a product's lifecycle. A key challenge is to develop further the procedures, processes and materials that will enable greater use to be made of WEEE. Legislation, largely being driven by Europe, is beginning to have an impact and there is now a well-defined move to adopt similar legislation in many other countries around the world. It will also be important that new environmentally related legislation, wherever it is implemented, does not differ substantially in scope from other similar legislation in a different region. Achieving a degree of harmonisation through an international standardisation process will become increasingly important as more environmental legislation is implemented in coming years. In attempting to achieve a harmonised approach to environmental regulation, particularly with legislation governing materials restrictions, there needs to be greater engagement between industry and policy makers so that legislators understand the environmental trade-offs inherent in materials substitution. Additionally, industry will need to become more proactive in negotiating on how the costs associated with sustainability best-practice and resource efficiency can be absorbed into the economy without threats to inflation and employment.

Although legislation can undoubtedly make a significant contribution in forcing the recycling of more materials from WEEE, there also needs to be a shift in individual thinking. This means a move away from the commoditisation of electronics that has occurred in recent years, to a situation where products have greater service lives before they are discarded and where refurbishment and re-use also have an enhanced role. There are also significant opportunities for material suppliers to play their part in developing new materials that will not only make it easier to treat electrical and electronic waste but which will also provide valuable recyclate sources for the manufacture of new products. Finally, the role of the designers must not be overlooked. Being at the beginning of the product lifecycle, they can have one of the biggest influences on what happens to products at end of life since they have control over many key factors such as material choices, quantities and specifications, as well as the product's likely service life and the ease of de-manufacturing at end of life. As will be seen throughout the rest of this book, what happens to waste electronics is increasingly influenced by considerations across the whole product lifecycle.

References and Further Reading

1. Greenpeace – Your Guide to Green Electronics http://www.greenpeace.org/international/news/green-electronics-guide-ewaste250806, accessed 03/04/08.

2. J. Pucket, S. Westervelt, R. Gutierrez and Y. Takamiya, *The Digital Dump – Exporting Reuse and Abuse to Africa,* The Basel Action Network, Seattle, WA, 2005.

3. DEFRA – Producer Responsibility http://www.defra.gov.uk/Environment/waste/topics/producer-responsibility.htm, accessed 03/04/08.

4. The Waste Electrical and Electronic Equipment (Waste Management Licensing) (England and Wales) Regulations 2006, Statutory Instrument 2006 No. 3289, available at http://www.opsi.gov.uk/si/si2006/20063289.htm#sch2, accessed 03/04/08.

5. Directive 2002/96/EC of the European Parliament and of the Council of 27 January 2003 on waste electrical and electronic equipment (WEEE) – Joint declaration of the European Parliament, the Council and the Commission relating to Article 9, Official Journal L 037 , 13/02/2003 P. 0024 – 0039, available at http://eur-lex.europa.eu/LexUriServ/LexUriServ.do?uri = CELEX:32002L0096:EN:HTML, accessed 03/04/08.

6. RoHS Regulations; Government Guidance Notes, URN 07/1234, SI 2006 No. 1463, July 2007, published by Department for Business, Enterprise & Regulatory Reform, London, available at http://www.berr.gov.uk/files/file40576.pdf, accessed 03/04/08.

7. Directive 2002/95/EC of the European Parliament and of the Council of 27 January 2003 on the restriction of the use of certain hazardous substances in electrical and electronic equipment Official Journal L 037, 13/02/2003 P. 0019 – 0023, available at http://eur-lex.europa.eu/LexUriServ/Lex UriServ.do?uri = CELEX:32002L0095:EN:HTML, accessed 03/04/08.

8. Directive 2005/32/EC of the European Parliament and of the Council of 6 July 2005 establishing a framework for the setting of ecodesign requirements for energy-using products and amending Council Directive 92/42/EEC and Directives 96/57/EC and 2000/55/EC of the European Parliament and of the Council, Official Journal of the European Union, L 191/45, 22 July 2005, available at http://eur-lex.europa.eu/LexUriServ/site/en/oj/2005/l_191/l_19120050722en00290058.pdf, accessed 03/04/08.

9. European Chemicals Agency, Annankatu 18, 00120 Helsinki, Finland, or visit http://echa.europa.eu/home_en.html, accessed 03/04/08.

10. DTI Global Watch Mission Report: Waste electrical and electronic equipment (WEEE): innovating novel recovery and recycling technologies in Japan, September 2005, published in January 2006 by PERA, Melton Mowbray, on behalf of the Department of Trade and Industry, available at http://www.oti.globalwatchonline.com/online_pdfs/36486MR.pdf, accessed 03/04/08.

11. Transposition of the WEEE and RoHS Directives in other EU Member States, November 2005, published by Perchards, St Albans, available at http://www.berr.gov.uk/files/file29925.pdf, accessed 03/04/08.

12. C. Cameron Beaumont, E. Bridgwater and G. Seabrook, *The National Assessment of WEEE Recycling Sites,* Future West and Network Recycling, 2004, available at http://www.networkrecycling.co.uk/downloadable-reports.htm, accessed 03/04/08, or contact enquiries@resourcefutures.co.uk.

13. WEEE Flows in London. An analysis of waste electrical and electronic equipment within the M25 from domestic and business sectors, September 2006, prepared by Axion Recycling Ltd, available at http://www.london remade.com/Uploads/medialibrary/weeeflowsinlondon.pdf, accessed 03/04/08.

14. NEC & UNITIKA Realize Bioplastic Reinforced with Kenaf Fiber for Mobile Phone Use, http://www.japancorp.net/Article.asp?Art_ID = 12143, accessed 03/04/08.

15. Vecap-Second Annual Progress Report 2007, published June 2007 by EBFRIP and BSEF, Brussels. Contact ebfrip@cefic.be.

Materials Used in Manufacturing Electrical and Electronic Products

GARY C. STEVENS AND MARTIN GOOSEY

1 Perspective

Any discussion of the materials used in electrical and electronic products requires consideration of the factors affecting materials choice and whole-life management. Today, primary environmental and sustainability-centred legislation arising in different regions of the world, and the economics of materials supply and management across the whole lifecycle, are primary drivers for the types of materials that will continue to be used in the future and those which need to be developed to replace those that cannot meet the requirements.

So, such a discussion must consider the requirements of current legislation and its implications for historically used materials which are now being returned for waste management at end-of-product life, and new materials which must satisfy current technical, economic and environmental acceptance criteria.

We begin by briefly examining the current legislative drivers in Europe, while acknowledging that similar drivers exist in other global regions. We then consider some of the primary materials currently used in electrical and electronic products, including some that will possibly be displaced by others in the future on the grounds of risk to human health and the environment.

2 Impact of Legislation on Materials Used in Electronics

2.1 Overview

With much legislation already in place that impacts the use of materials in electronics applications, it is likely that this trend will continue, with many

Issues in Environmental Science and Technology, 27
Electronic Waste Management
Edited by R.E. Hester and R.M. Harrison
© Royal Society of Chemistry 2009
Published by the Royal Society of Chemistry, www.rsc.org

more commonly used materials and additives being proscribed. This could come from a strengthening of the existing legislation, such as the RoHS Directive (the Restriction of the use of certain Hazardous Substances in electrical and electronic equipment), as well as from additional legislation. For example, the REACH Regulations, with their requirement for Registration, Evaluation and Authorisation of Chemicals, is likely to lead to the demise of some important chemicals and consequent changes in the formulation of many materials used in electrical and electronic applications. The Energy-using Products Directive (EuP) will require the electronics industry to take a more holistic approach to the way it manufactures its products, with emphasis being placed on all aspects of a product's lifecycle from eco-design to end of life.

The encouragement of eco-design principles will lead to the integration of environmental considerations during the design and materials-selection phases of a product. The European Parliament and the Council adopted a final text for the EuP Directive 2005/32/EC in July 2005. Actual measures have been decided on a product-by-product basis under the supervision of a designated panel of EU member state experts as part of the so-called 'fast-track comitology procedure'. Priority products include heating, electric motors, lighting and domestic appliances. Ultimately, this framework directive will cover all products consuming energy, apart from motor vehicles, and it is thought that these could account for 40% of the carbon dioxide emissions responsible for global warming, which are to be reduced under the Kyoto Protocol.

As this type of legislation continues to be implemented, the electronics industry will need to keep pace with the changes that will be required and make materials selections and substitutions of older, and in some cases proscribed, materials that satisfy legislative requirements. Some of the issues with which the industry will need to come to terms are:

- restrictions and limitations on the types of materials that can be used in electronics
- end-of-life recovery, re-use and recycling requirements for materials including selective materials recycling
- enhancements in the energy efficiency of products, in terms of both the manufacturing and use phases
- adoption of a more holistic approach, embracing all aspects of the product lifecycle from design through to end of life, including recycling and re-use
- the adoption of more sustainable business practices which will include materials sourcing and improved materials stewardship along the supply chain
- the development of new materials that improve product performance and offer greater ease of recycling and re-use

For many large multi-national companies one of the key challenges will be the integration of cost-efficient materials compliance strategies across multi-tiered global supply chains. With global supply chains often emanating from the Far East there will clearly be an increasing need to minimise the level of supply chain confusion that could emerge as a result of the introduction of disparate legislation in different global regions, countries and national regions or states. There are already issues about the supply of RoHS-compliant components for certain regions and not others. Ideally, the industry should be seeking to achieve some degree of consensus with regard to environmental legislation since one clear benefit would be a reduction in the need to manufacture 'region specific' products. It is important that new environmentally related legislation, wherever it is implemented, does not differ substantially in scope from other similar legislation in a different region. Achieving a degree of harmonisation through an international standardisation process will become increasingly important as more environmental legislation is implemented in coming years.

Although there is much producer responsibility legislation that has an impact on the materials used in electronics and what happens at end of life, one of the most important is the RoHS Directive and this is covered in more detail in the following section.

2.2   The RoHS Directive and Proscribed Materials

The RoHS Directive proscribes substances that are hazardous and that pose serious environmental problems during the disposal and recycling of WEEE. In line with the similar requirements of the End-of-Life Vehicles Directive, the targeted substances are the heavy metals mercury, lead, cadmium and hexavalent chromium, as well as polybrominated biphenyls (PBBs) and two polybrominated diphenyl ethers (PBDEs) (penta-PBDE and octa-PBDE). The substances are clearly defined and the levels above which they must not be present are specified. Electrical and electronic products covered by the WEEE Directive (except medical devices and monitoring and control instruments) and put on the market after 1 July 2006 cannot contain any of the materials specified in Table 1 at concentrations higher than the maximum permitted level.

The maximum permitted level is not as a percentage of the total weight of a product but as a percentage weight of a 'homogenous material'. As the RoHS Directive evolved there was much discussion about what exactly constituted a homogenous material. However, the wording of the Directive states:

'A maximum concentration value of 0.1% by weight in homogeneous materials for lead, mercury, hexavalent chromium, polybrominated biphenyls (PBB) and polybrominated diphenyl ethers (PBDE) and of 0.01% weight in homogeneous materials for cadmium shall be tolerated. Homogeneous material means a unit that cannot be mechanically disjointed in single materials.'

The UK's Department of Trade and Industry (DTI) (now The Department for Business, Enterprise & Regulatory Reform, BERR) has attempted to give some definition to the terms and described what is meant as follows:

- 'homogeneous' means 'of uniform composition throughout'
- Examples of 'homogeneous materials' therefore include individual types of plastics, ceramics, glass, metals, alloys, paper, board, resins and coatings.

Table 1 Materials targeted by the RoHS Directive.

Material	*Maximum permitted level*
Lead	0.1%
Mercury	0.1%
Hexavalent chromium	0.1%
Cadmium	0.01%
Polybrominated biphenyls	0.1%
Pentabromodiphenyl ether	0.1%
Octabromodiphenyl ether	0.1%

- In this context the phrase 'cannot be mechanically disjointed' is also understood to mean that the materials are not able to be separated further by mechanical actions such as unscrewing, cutting, crushing, grinding and abrasive processes.

Although the RoHS Directive is European legislation, its implementation also has supply-chain ramifications that are global. For example, leading electronics manufacturers in Japan have, for several years, been preparing for the Directive's implementation. Ricoh has required its suppliers to 'attest to the non-use of banned substances' since July 2002, when the company made a revision to its green procurement standards. Other significant electronics manufacturers such as Matsushita, Oki and Sony also oblige suppliers to submit similar certifications and Canon and Hitachi even launched divisions dedicated to preparation for the impact of the RoHS Directive.

While it is quite clear in some applications where RoHS-proscribed materials may occur, *e.g.* lead in traditional solders, it is also often very difficult to determine if materials and components used in electronic products are indeed RoHS compliant. This is especially challenging in an industry that has long and complex international supply chains and where a large proportion of the components used are sourced from overseas. A key part of achieving RoHS compliance involves having a good understanding of where proscribed materials may occur and the following section gives some examples. From an end-of-life perspective, it is also vitally important that the compositions of materials appearing for end-of-life recycling are known, especially the presence of RoHS-proscribed materials. If materials are to be recovered and re-used in applications where RoHS compliance is required, there must be no possibility of introducing non-compliant materials from the recyclate.

Although the RoHS Directive is a piece of European legislation, there is similar legislation being implemented in other parts of the world. A key recent example is the Chinese version of RoHS (Management Methods for Controlling Pollution by Electronic Information Products), which, although addressing the same materials as detailed in the European RoHS, has a completely different approach to implementation.

3   Where do RoHS Proscribed Materials Occur?

It is important for all parts of the supply chain to understand where RoHS-proscribed materials may be encountered and with what they should be replaced. Although much of the global electronics supply chain has already become RoHS compliant, there are still many components and materials produced that contain the proscribed materials. The inadvertent use of components containing these materials would subsequently render the final products non-RoHS compliant and leave the producer open to the possibility of having to remove products from the European market. The introduction of proscribed materials into waste streams used for recycling can also compromise the ability to re-use such materials in applications requiring RoHS compliance.

3.1   Lead

The proscription of lead has caused significant concerns for electronics manufacturers and the need to replace it before July 2006 gave the industry a huge task in terms of evaluating, testing and qualifying potential substitutes, particularly in soldering applications. Although there are numerous 'lead-free' solders available, there are many issues that need to be addressed before they can be successfully implemented as viable alternatives. These issues include, amongst others, the compatibility of new solders with printed circuit board and component finishes, the ability of existing materials and equipment to handle higher soldering temperatures and the selection and supply of components that can survive higher soldering temperatures and give long-term reliability.

Lead can also be found in other components, products and parts of the electronics assembly process. A good example is in PCB manufacturing where lead was once widely used as a solderable final finish. The key will be to know what finish is used, since it is important that the combination of solderable finish, solder and component finish are compatible in order to avoid solderability and subsequent reliability issues. Lead is also used in the solderable finishes of component connectors. A challenge for electronics assemblers is the identification of the finish on component leads and terminations since there is, as yet, no agreed standard convention for identifying whether or not a component finish is lead-free. Given that components with and without lead-free finishes may appear visually identical, it is clear that careful attention needs to be paid to component sourcing, selection and storage, *etc*. Also, from an end-of-life and recycling perspective, the presence in waste streams of a mixture of lead-containing and lead-free electronics is likely to complicate the recycling process for many years to come.

3.2   Brominated Flame Retardants

Brominated flame retardants, in isolation or in combination with antimony trioxide, perform a valuable function in preventing fires and it is important to note, therefore, that the RoHS Directive does not proscribe the use of all brominated flame retardants but just two particular types. These are poly-brominated biphenyls and two examples of the polybrominated diphenyl ethers (pentabromodiphenyl ether and octabromodiphenyl ether). Brominated flame retardants are widely used in electronics products; for example, they are found in PCBs, semiconductor encapsulants, cables and connectors, as well as in the polymers used in equipment housings and enclosures as discussed below. Reactive brominated flame retardants, such as the tetrabromobisphenol A-based compounds used in most PCB laminates and semiconductor encapsulants are not currently addressed by this legislation. However, suppliers of some polymers have already commercialised new bromine-free materials and many companies have already begun to avoid the use of brominated flame retardants in their products. It is also worth noting that pentabromodiphenyl ether is reportedly still used in a small percentage of Far East produced FR2 printed circuit board laminates.

3.3 Cadmium, Mercury and Hexavalent Chromium

These three metals can occur in a wide range of applications in electronics and in this respect they are perhaps more troublesome from a compliance perspective since there are so many potential, and sometimes unexpected, applications. Cadmium has been widely used in specialist electroplated parts but it is most likely to be found in electronics applications in nickel-cadmium batteries. In most cases, cadmium plating is used for applications that are not covered by the RoHS Directive. However, items or components containing cadmium must be removed from end-of-life electronics before disposal. Cadmium is also found in cadmium sulfide-based photodetectors and as a component of the green and blue phosphors used in older colour television cathode ray tubes. Importantly, it has sometimes been found in pigments used to colour plastics and can thus be found in various electronic products. For example, pigmented ABS plastics have been widely used in items such as telephones and electricity cables. Cadmium has also been found in certain electrical and electronic components such as surface mount device chip resistors, infrared detectors and semiconductor chips. Cadmium Mercury Telluride is also used in infrared detectors which have unique characteristics.

It has been estimated that 22% of all mercury is employed in electrical and electronic equipment. It is used in thermostats, sensors, relays, switches, medical equipment, fluorescent lamps, mobile phones and batteries. Mercury is also used in the backlights of flat-panel displays and this usage has increased as liquid crystal displays increasingly replace conventional cathode ray tubes. Laptop computers, flat-panel displays and digital cameras may all, therefore, contain small amounts of mercury. Use in batteries, however, has been decreasing for some years and it is now found only in button cells that power relatively small electronic goods such as watches, toys and cameras.

Hexavalent chromium compounds are widely used in electroplating and metal treatment processing. They will passivate the surface of zinc and zinc alloy electrodeposits with a thin film that provides end-user benefits such as colour, abrasion resistance and increased corrosion protection. Hexavalent chromium, although not widely used directly in the electronic components, is thus often found as a coating on the various brackets, fittings and other metal parts used in many actual products. It may be used to provide corrosion-protection to steel parts, as well as screws and nuts, and thus may be found in both consumer goods and those destined for operation in harsher environments. It also finds use as an anti-corrosion material for the protection of carbon steel cooling systems in absorption refrigerators. The difficulty for producers responsible for end-of-life electronics is that, for some large products with a significant non-electronic content, there may well be individual component parts that have been plated or treated in some way that means they contain hexavalent chromium. The challenge will be in identifying which, if any components, these are. To a similar extent this also applies to cadmium.

4   Soldering and the Move to Lead-free Assembly

4.1   Introduction

The development of the RoHS Directive has raised many concerns, especially with respect to the fact that lead has effectively been proscribed in most electronics goods put on sale in the European market since July 2006. Fortunately, much work has been carried out to define viable lead-free solders, material combinations and process conditions that enable the production of lead-free assemblies in good yields and with reliability equal to or better than those based on the tin-lead solders they replace. There are numerous lead-free solder alloys commercially available with a range of melting points/ranges, but those that are proving most popular for both reflow and wave soldering within Europe have melting points significantly higher than the tin-lead solders they replace. These higher melting points have important ramifications for the materials and components being assembled because processing temperatures will have to increase. Coupled with the fact that lead-free solders behave somewhat differently from tin-lead solders, there are a number of important considerations that must be addressed if successful lead-free assembly is to be achieved. This section outlines implications of moving to lead-free in the context of both electronics assembly and the broader lifecycle considerations including end of life and recycling.

4.2   Lead-free Solder Choices

Traditional electronic assembly typically relies on the use of tin-lead-based solders with tin : lead compositional ratios of 60:40 or 63:37 that melt at, or around, 183 °C. These are ideal for electronics assembly in that the soldering temperature is high enough to enable product operation over a relatively wide temperature range without requiring a soldering temperature that can damage

Table 2 Examples of tin-silver-copper and related lead-free solder alloys.

Alloy type	Composition	Melting Point/°C^a
Tin-silver-copper	Sn96.5/Ag3.0/Cu0.5	219.8
Tin-silver-copper	Sn95.5/Ag3.8/Cu0.7	218.8
Tin-silver-copper	Sn95.5/Ag4.0/Cu0.5	220.2
Tin-silver-copper-antimony	Sn96.2/Ag2.5/Cu0.8/Sb0.5	217.0
Tin-copper	Sn99.3/Sn0.7	227.0

^{a}Melting points may vary slightly depending on measurement conditions.

components and circuit boards. There are many lead-free solder systems that have been proposed as replacements for conventional lead-based solders and they have melting ranges from much lower than conventional solders to much higher. In order to bring some degree of conformity to lead-free assembly, various organisations (*e.g.* Soldertec, IPC and Intellect) have recommended that the industry minimises its choice of solders and specific alloy types that are suitable for most applications have been defined.

Consequently, solders based on tin-silver-copper alloy compositions are widely used for both reflow and wave soldering and tin-copper alloys are also used for wave soldering. The tin-silver-copper solders, which are also known as SAC alloys, have compositions with melting ranges between 215 and 220 °C. For wave soldering, the tin-copper eutectic alloy (Sn-0.7Cu) melts at 227 °C and this represents one of the lowest-cost lead-free alloys available, although SAC alloys are also widely used in wave soldering. Thus, for most European companies switching to lead-free solder, the key challenges throughout the whole manufacturing process are really focused on the use of different solder compositions and the fact that they will, in all probability, require significantly higher processing temperatures.

These higher soldering temperatures can have a significant impact on materials, components and processes and it is vital that soldering operations are modified accordingly if successful soldering is to be achieved. Examples of the tin-silver-copper and tin-copper alloys are given in Table 2.

5 Printed Circuit Board Materials

5.1 Introduction

One of the key requirements in all electrical and electronic devices is the need to provide convenient, efficient and low-cost interconnections between the various components, displays and other devices in order to enable them to function. Since the beginning of the semiconductor age this has been achieved in most electronic products through the use of printed circuit boards (PCBs). Whilst often appearing to be no more than a series of copper conductor patterns on an insulating substrate, PCBs are often very complex multilayer devices that must function with absolute reliability in a wide range of environments.

Consequently, there are many materials-related challenges that must be addressed in order to produce circuit boards that meet these requirements and which provide the necessary electrical functionality. It is also worth noting that the PCB industry uses materials on a vast scale: the value of global PCB production for 2007 has been estimated at just under $50 billion.

Until recently, most development work on PCBs and their materials had been focused on meeting the incessant demands of the electronics industry for lower cost, improved performance and increased functionality within a smaller space. Thus, the focus had been on producing circuit boards that were cheaper to manufacture, which had more interconnections per unit area and which could also operate at increasingly higher frequencies. This has been and continues to be achieved through both incremental development of existing materials and processes and the development of completely new approaches. While the PCB industry and its materials and process suppliers have been very successful in meeting the performance criteria set by the electronics manufacturers through the development of these new materials and processes, it is true to say that little, if any, consideration has been given to end of life and recycling considerations for both bare and assembled circuit boards. The growing need for the electronics industry to operate more sustainably, due to both societal and legislative pressure, is forcing all parts of the electronics industry to adopt a more holistic approach and to consider what happens when products are discarded. This is nowhere more true than in the PCB industry where a variety of materials are used that have major implications in the context of end-of-life disposal and recycling. The proliferation of producer responsibility and related legislation has also had a significant impact on the PCB industry with, for example, the RoHS Directive, causing major changes in the way that circuit boards are assembled and having a direct impact on the materials that are used. More specifically, the move to lead-free assembly detailed elsewhere in this section has necessitated significant materials-related changes in PCBs and thus there are ramifications for end-of-life and waste treatment practices. This section outlines the materials that are used in PCBs in the context of a more holistic approach to electronics manufacturing that must increasingly embrace end of life, waste management and recycling requirements.

5.2   PCB Materials

PCBs are typically manufactured using copper-clad glass-fibre-reinforced laminate materials in which the supporting insulator between the conductors is made from a reinforced polymer. For electronic products that require only a single layer of copper interconnection, the supporting dielectric material is often made from low-cost materials such as paper-reinforced phenolic resins. While these types of laminate are widely used to produce large volumes of boards for consumer products such as radios, calculators and televisions, there are also many applications that require much more complex multilayer interconnect structures and better performance than these low-cost materials can deliver. The multilayer boards that would, for example, be found in mobile phones and personal computers have traditionally been made from laminates based on epoxide chemistry and the most widely used material is known as FR4. FR4 is a glass-fibre-reinforced laminate that employs an epoxide resin system based on the diglycidyl ether of bisphenol A. FR4 laminates are also required to meet various flame retardancy performance standards (*e.g.* UL94-V0) and, as such, also contain appreciable quantities of brominated resins to impart the required degree of flame retardancy. As with the paper-reinforced phenolic laminates, FR4-type laminates find extensive use in many applications and will continue to do so for the foreseeable future.

Traditionally, FR4 laminates made using bisphenol-based epoxy resins had glass transition temperatures in the region of 130 °C, but in recent years these values have increased gradually to around 150 °C or even more. However, the legislation-driven move to lead-free assembly, where higher-melting-point solders are used, and the possibility of boards being exposed to multiple soldering operations, has driven the requirement for the thermal performance of laminate materials to be improved further. Laminates now often need to be able to survive soldering temperatures that could be as high as 260 °C. In order to achieve improved thermal stability, higher-T_g materials (the glass transition temperature (T_g) of a non-crystalline material is the critical temperature at which the material changes its behaviour from being 'glassy' to being 'rubbery') have been developed that have T_g values stretching well above 200 °C and in some special cases up to nearly 300 °C. Examples of these laminates include, amongst others, those based on polyimides, cyanate esters, allylated poly-phenylene ethers and the so-called BT-epoxy and tetrafunctional epoxy systems as illustrated in Table 3.

In addition to thermal stability, it is increasingly important, as device features get smaller and interconnect densities increase, that thermal stress issues

Table 3   Properties of some high-performance laminates compared to FR4.

Laminate Material	T_g (°C)	Dielectric constant (10 GHz)	Dissipation factor (10 GHz)	Relative price
Standard T_g epoxy	130–150	4.5	0.022	1
High T_g epoxy	170–180	4.4	0.02	1.15–1.25
Polyphenylene ether	175	3.4	0.009	3
Epoxy/polyphenylene oxide	180	3.9	0.013	1.7
Bismaleimidetriazine	180	4.1	0.013	2 to 3
Epoxy/cyanate ester	210	3.6	0.014	
Cyanate ester	240	3.8	0.009	2.5–4
Polyimide	280	4.3	0.02	2.8–4.5
Hydrocarbon/ceramic	<280	3.48	0.004	
Liquid crystal polymer	280	2.8	0.002	

Note: information taken from laminate manufacturers' literature.

are also minimised. Thermal stresses occur because the organic components of a laminate have much higher thermal expansion coefficients than the copper metallisation to which they are required to form a strong and stable bond. Above the T_g of the laminate resin, the rate of thermal expansion increases and, therefore, lower-T_g materials exhibit a greater degree of expansion as they are heated to elevated temperatures for soldering. The use of woven glass-fibre reinforcement in the laminates helps to constrain these expansion effects and to minimise the effects of the thermal expansion coefficient mismatch. However, movement in the vertical (z) axis of a board is most significant and can result in interconnection failures. Therefore, by increasing the T_g of the laminate to nearer the temperatures likely to be encountered during assembly, it is possible to minimise the overall expansion and thus reduce the influence of thermal stress.

It is important to note that, while enhanced T_gs are one requirement of an advanced laminate, another increasingly important requirement is for enhanced dielectric properties. As electronic-product performance continues to be enhanced, devices are operating at much higher frequencies than previously and, with frequencies for many applications often being in the gigahertz range, there is a growing need to produce laminates which do not compromise signal transmission speeds and integrity. The two key properties of a laminate that need to be improved are the relative permittivity (dielectric constant, D_k) and the dissipation factor/loss (D_f). Signal propagation speed is inversely proportional to the square root of the dielectric constant and, consequently, a lower D_k permits faster signal speeds. Signal loss through dissipation directly relates to signal speed and thus, the lower the dissipation factor, the greater the efficiency of signal propagation.

While FR4 epoxide-based laminates have reasonable dielectric properties for many applications, there is a requirement to reduce their values for higher-frequency applications. Epoxide resins are relatively polar materials and FR4-type laminates typically have dielectric constants of around 4.5 and dissipation factors of 0.02 (depending on the frequency). A significant contribution to this value comes from the large quantity of woven glass-fibre that is used. Some of the newer materials mentioned above, especially when used with new types of reinforcement, offer the promise of both higher T_g values and enhanced dielectric properties. Typically, however, they tend to cost much more than FR4. The properties of some example laminate types are shown in Table 3.

5.3 Provision of Flame Retardancy in PCBs

PCB laminates are required to meet certain flammability requirements such as those specified by the Underwriter's Laboratory in UL94-V0 and thus they have traditionally been formulated to contain flame- retardant components. In the case of FR4-type epoxide-based materials, this flame retardancy is achieved through the incorporation of brominated resins into the polymer matrix. By far the most widely used materials for this application are those based on tetra-bromobisphenol A-based resins (TBBPA). These resins perform very well, but in recent years there have been many negative reports about the impact of brominated species on the environment and, in particular, their persistence in the environment. Brominated flame retardants were candidates for proscription in early iterations of the WEEE Directive but the pertinent legislation is now in the RoHS Directive. At one point TBBPA-based resins were specifically cited as a class of brominated flame-retardants that would be proscribed, but the legislation has evolved to become more specific and, at the time of writing, the flame retardants detailed for proscription belong to certain members of the non-reactive family of polybrominated diphenyl ethers and the polybrominated biphenyls. It is important to appreciate that reactive FRs such as TBBPA become covalently bound into the resin matrix and they are unable to migrate or be released as the same free compound once did. Therefore any concerns about TBBPA reside in the original production and handling of the compound up to the point of incorporation into the resin. TBBPA is not specifically cited for proscription at this time, although the situation could change in the future if further studies of its impact reveal a problem.

The attention brought to the environmental issues associated with brominated flame retardants and the fear that TBBPA-type resins might also be banned has prompted laminate manufacturers to develop alternative flame-retardant systems. As with other environmental issues pertinent to the electronics industry, Japanese companies have been at the forefront of technical activities to develop more environmentally friendly laminates, although it is fair to state that there has also been an appreciable amount of development work undertaken by both European and American companies. There are various approaches that have been adopted to provide viable bromine-free flame retardancy to laminates. These include the use of phosphorus, in various forms, as well as antimony oxide, hydrated metal oxides and nitrogen-containing organics. Hitachi has, for example, developed a laminate system based on the use of a new resin system containing a large quantity of nitrogen and a high inorganic-filler content. Whilst many of these alternatives may offer flame retardancy properties, some are not particularly desirable as alternatives to bromine when considered from either an environmental or a health and safety perspective.

Phosphorus has been used in various forms to provide flame-retardant properties to laminates, although it acts *via* a totally different mechanism from the brominated systems. It works in the solid phase by producing char that stops the propagation of fire and in some cases its effectiveness is enhanced by the presence of nitrogen. Phosphorus is toxic when burnt and it is also more expensive to use than bromine. There have also been suggestions that free phosphorus can lead to the poisoning of plating baths, although specific references to this effect are illusive. Phosphorus-containing laminates are also more sensitive to moisture absorption and thus there is the possibility of a laminate's electrical properties being degraded. Some laminate manufacturers claim to have overcome the issues associated with phosphorus by using new resins that have chemically bonded phosphorus (and sometimes nitrogen as well) in their polymer structures. The laminate manufacturer, Isola, has produced a base material similar to FR-4 that is known as DURAVER. The resin matrix is based on a phosphorus-modified epoxy resin and conventional glass fabric is used for reinforcement. The material meets the requirements of UL94-V0 without the need for the addition of antimony compounds.

There are some important issues around whether laminates have to be bromine-free or, as is often quoted, halogen-free. The specific concerns from a legislation perspective have been about the use of brominated flame-retardants but it is important to remember that epoxy resins often contain low levels of chlorine, which is also a halogen, as impurities. Also, some of the best-performing high-frequency laminates are based on fluorine-containing polymers. There have also been questions raised over at exactly what level of halogens a laminate is considered to be halogen free. It seems that the broadly accepted upper limit for bromine and chlorine in a halogen-free system is 900 ppm or 0.09%. This figure originated in the Far East and has been adopted by a number of major companies. The figure of 900 ppm is around two orders of magnitude lower than would typically be found in a brominated flame-retarded laminate.

Thus, it can be seen that, in addition to the need for enhanced performance properties in laminates, the emergence of new environmental legislation is also driving the need for considerable changes in PCB manufacturing processes and particularly with respect to laminate materials. It is likely that over the next few years there will be further considerable progress made in the development of new laminate materials that are bromine-free, have higher thermal stabilities, have better dielectric properties and which are more amenable to treatment and recovery at end of life. Technically, solutions exist to meet these requirements but one of the biggest challenges will be to produce these new products at an acceptable cost.

Currently, there is no accepted technology for the recovery of the constituent materials of PCB laminates, although there have been some attempts to use the materials in particulate form as a filler in other epoxy composites. An example is that taken from the work of one of the authors (G. C. Stevens *et al.*; Polymer Research Centre report – obtainable from GnoSys UK) in using particulated PCB laminate with dual microwave-thermal curing of castings and mouldings using conventional epoxy resins with appropriate compatibilisers.

5.4   Non-ferrous and Precious Metals

PCBs are populated with a wide variety of electronic components that contain a wide variety of precious metals including gold, silver, copper, platinum, rhodium and palladium. These are present in small quantities but it can be economic to apply wholesale fragmenting, metal dissolution and recovery technologies to recover significant amounts of these metals. The primary methods used are smelting and electrochemical recovery from dissolved metal solutions.

6 Encapsulants of Electronic Components

Modern electronic products utilise components and circuitry that contain very fine features and interconnections which make them susceptible to damage during handling and from exposure to typical operational conditions. Consequently, there is a need to provide protection to these delicate devices at the semiconductor surface, device and assembled-board levels. This is typically achieved by using a wide range of encapsulant materials, which have evolved over many years. These are based on a number of well-established basic polymer chemistries and they are applied using techniques such as transfer moulding, powder coating, liquid injection moulding, spraying and potting.

Although there is a wide range of potential encapsulant chemistries that have been used, the predominant ones are the epoxides and silicones, with polyurethanes and acrylates also finding use. In the case of discrete semiconductor devices, these are typically encapsulated using highly filled epoxide-based products using a transfer-moulding process. Transfer-moulding involves the rapid heating and liquefication of the moulding compound, after which it is transferred into a multi-cavity mould containing large numbers of the devices to be encapsulated. At the temperatures used, the epoxide undergoes a rapid curing reaction and solidifies around the devices to provide the required physical form and a protective barrier. After removal from the mould, and subsequent mechanical operations, the encapsulated device is in its final form and ready for assembly. The epoxides used in this type of encapsulant are typically based on novolacs and the moulding compounds also employ additional novolac resins to give a highly cross-linked, high-T_g, cured product. Rapid reaction is achieved by the use of proprietary catalysts and relatively high temperatures. In addition to the basic reactive ingredients, these moulding compounds also contain a large number of other ingredients including brominated flame retardants, usually in combination with antimony oxide, mould-release agents, flow modifiers and a high loading of an inorganic filler such as silica.

Although the encapsulation of semiconductor devices by transfer moulding has been the predominant method used for protecting individual semiconductor devices such as conventional integrated circuits, other discrete components often also need to be encapsulated. Although this can sometimes be achieved using transfer moulding, other techniques such as dip coating or powder coating are widely used. In the case of dip coating, the encapsulant will typically be a liquid system which utilises either two components that are mixed just prior to use, and which then react very quickly once mixed, or a single component system with a latent catalyst that is activated by heat, *e.g.* from a pre-heated component or from a subsequent heating stage. Powder coating has been used for components such as tantalum capacitors and in this case the basic chemistry and formulation is very similar to that employed for transfer moulding, with modifications made to suit the different application method.

In addition to the encapsulation of discrete components, encapsulation is also used to provide protection to more complex modular assemblies where there may be several semiconductor devices assembled on a substrate. This type of assembled device was originally known as a multichip module but there have been many iterations in recent years, including the currently popular System in Package approach. There has been a proliferation of different types of semi-conductor assemblies that need to be encapsulated and the materials and encapsulation methods have had to evolve or be developed to meet these needs. Encapsulants and the equipment used to dispense them have thus become a fast-growing and sophisticated area of semiconductor packaging that has led to, for example, the use of vacuum dispensing of encapsulants for chip-scale packaging.

Once individual devices have been encapsulated, they are then assembled onto a circuit board to give the final product and quite often there is a need to provide a protective encapsulating coating to this complete assembly. Board-level encapsulation can be achieved by dip or spray coating and a wide variety of materials are suitable including silicones and polyurethanes. Silicones typi-cally provide outstanding performance, but there is often a trade-off between performance and cost and for less-demanding applications alternative materials may be preferable.

While encapsulants enable current electronic devices to survive in relatively hostile environments, they can also make access to individual components on a board difficult. This not only makes repairs and replacement of faulty com-ponents more difficult but it also makes access to components for recovery and re-use at end of life more difficult as well.

Currently, no encapsulant materials are recovered or recycled as both their chemistry and physical presence are not amenable to recycling and such an activity would be uneconomic.

7   Indium Tin Oxide and LCD Screens

The development of liquid crystal displays (LCDs) requires the introduction and then growing use of transparent conductive materials. The need to apply an electric field across the liquid crystal materials in a display means that the transparent faces of these displays require an electrically conductive coating. This is typically achieved using a physically deposited layer of indium tin oxide (ITO). Although, these coatings are typically very thin, the rapid replacement of conventional cathode-ray-tube-based televisions with those using LCDs, and the fact that LCDs are increasingly used as the preferred type of display medium in many types of electronic products, means that the quantities of ITO used are growing significantly and it is likely that ITO supply may not be able to keep pace. Recycling will become imperative if viable alternatives cannot be found.

Increasing numbers of LCDs are already reaching end of life and valuable materials such as indium are entering the electronics waste stream. Waste electrical and electronic equipment containing LCDs has been identified as one of the fastest growing sources of waste in the EU, with conservative estimates showing the volume increasing by between 16 and 28% every five years. Up to now, LCDs have typically been disposed of at end of life using conventional disposal routes, with no attempt being made to recover the valuable materials, such as ITO, that they contain. Flat panel displays use $\sim 80\%$ of global indium production and it has even been predicted that supplies will be exhausted by no later than 2025. This is clearly not sustainable and represents a waste of valuable resources, as well as a lost opportunity for recyclers. With the drive to more sustainable approaches to electronic product manufacturing and recycling, and the growing volumes of LCDs that are already entering the waste stream, there is clearly a need for new processes that enable valuable materials such as indium to be recovered and recycled from end-of-life or faulty LCDs. There have been some reports of new approaches being developed for the treatment of end-of-life liquid crystal displays in order to affect material recovery, but commercial implementation remains limited.

The Japanese electronics company Sharp has, for example, patented an apparatus and method for treating waste LCD panels.[1] This patent describes a multistage process employing both mechanical and chemical methods for recovering a variety of materials from LCDs. LCD panels are first crushed into glass cullet (small chips) and the indium on the surface of the glass is then dissolved using an acid solution. It is claimed to be a simple process using common chemicals that eliminate the need for a large energy input to create high temperatures or high pressures. The indium metal is recovered as the hydroxide and is claimed to be of high purity.

Within LCD screens a significant amount of high-quality thin-section glass, aluminium foils and polarised plastic films are used, all of which have economic value if they can be recovered and concentrated as single materials streams rather than as mixed waste streams. More effort is required to develop effective and economic separation technologies for these more complex products.

Further information on the recycling of liquid crystal materials and other materials from LCD screens can be found elsewhere in this book.

8 Polymeric Materials in Enclosures, Casings and Panels

Many electrical and electronic products contain metal structural panels and enclosures and casings. Both ferrous and non-ferrous metals can be easily identified, recovered and recycled using conventional metals recycling technologies. However, in the area of white goods and consumer electronics there is significant use of engineering thermoplastics.

8.1 Product-related Plastic Content

The Waste from Electrical and Electronic Equipment (WEEE) Directive (2002/96/EC) introduces, amongst other things, a category listing of the various types of WEEE with differing recycling and recovery rates (see Table 4).

The categories have been introduced to assist the segregation of product types into groups that are amenable to common recycling methods and where the recovery and recycling rates are considered achievable. A list of these rates

Table 4 WEEE categories and their proposed recovery and recycling rates.

WEEE category	Recovery rate	Recycling rate
1,10	80%	75%
3,4	75%	65%
2,5,6,7,9	70%	50%

for each category is provided in Table 4 and each category is described in more detail below:

Category 1 – Large Household Appliances

This group contains large items such as refrigerators, freezers, fridge freezers, washing machines, dishwashers, dryers and cookers. It also includes smaller items such as microwave ovens, heaters, radiators, fans and air conditioning units.

Category 2 – Small Household Appliances

This category is composed of items such as vacuum cleaners, irons, toasters, fryers, kettles, scales and other domestic items such as hair care and timers.

Category 3 – IT and Telecomms Equipment

Items including computers, mice, keyboards, printers, copiers, faxes, telephones and CRT and flat-screen monitors fall within this grouping.

Category 4 – Consumer Equipment

Items including TVs, radios, DVD players, VCRs, CD players, Hi-Fi items, speakers, amplifiers and musical instruments fall within this group.

Category 5 – Lighting

The category includes lighting units for both fluorescent bulbs and element-based light sources. None of the units in the trial had the actual bulb or tube in place. This category excludes household luminaries.

Category 6 – Electrical and Electronic Tools

Drills, saws, equipment for turning, milling, sanding, grinding, cutting, shearing *etc.*, sewing machines, lawnmowers and strimmers fall within this category.

Category 7 – Toys and Leisure

These items include exercise equipment through to computer game consoles.

Category 8 – Medical Devices

Specialist waste, medical devices, except implanted and infected products (*e.g.* ventilators, analysers).

Category 9 – Monitoring Equipment

Generally limited to equipment such as flow gauges and measuring equipment.

Category 10 – Automatic Dispensers

Fairly rare waste such as drinks dispensers, chocolate dispensers, ATMs.

A possible route for mixed WEEE is for it to be separated into its categories so that it can be tracked and estimates made of the materials content and composition before being treated.

Although there are a number of figures quoted for plastic content in WEEE (22% by weight, ICER[2]) and the breakdown into type of plastic consumed in EE production, this really only acts to give an overview of all WEEE and the potential for plastics recovery. With recyclers and reprocessors likely to specialise in certain product types and components, to simplify procedures and increase efficiency, it makes sense to have a look at the category compositions.

A recent trial was carried out and reported by DEFRA to determine the make-up of WEEE arriving from Civic Amenity (CA) sites.[3] The WEEE was classed as small mixed WEEE (SMW) since certain components of WEEE were not present as they are generally separately treated. These are:

1. Refrigeration units, which under EC directives must be properly dismantled by authorised treatment facilities to remove any Ozone Depleting Substances (ODS).
2. Large white household items – the size and weight of these items means that they are normally dealt with by another route, for example, alongside end-of-life vehicles (ELV) in mass shredding plants.
3. Cathode ray tube (CRT) displays – these items must be dismantled separately under health and safety regulations.

The breakdown of the categories of small mixed WEEE is shown in Figure 1. This shows that with large white goods such as refrigerators and washing machines removed, and the removal of CRT display units, all of which in theory will be dealt with separately, the majority (65%) is contained in

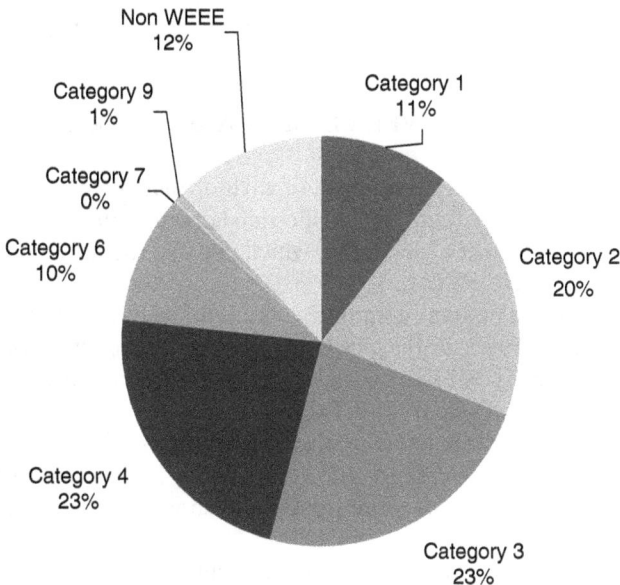

Figure 1 Breakdown of small mixed WEEE into category.

Table 5　Breakdown of composition of Categories (PCB – Printed Circuit Boards, Other – majority is wood or glass) by weight %.

Category	Metal (%)	Plastic (%)	PCB (%)	Other (%)
Category 1	76.7	13.8	0.4	9.1
Category 2	38.2	59	0.2	2.7
Category 3	59.9	33.3	4.6	2.2
Category 4	53.5	26.5	4.6	15.4
Category 6	55.3	41.9		2.9
Fridges	69.5	14.6		15.9
Large white goods	61.6	12.4		25.9
CRT monitors	13.4	20.9	13.5	52.2
CRT TVs	15.1	16.7	9.7	58.6

Table 6　Plastic composition by type, for various WEEE (weight %) (MOEA/MBA/APC 2000).

Plastic resin	Television plastics (%)	Computer plastics (%)	Miscellaneous plastics (%)	% of total sample
HIPS	82	25	22	56
ABS	5	39	41	20
PPO	7	17	4	11
PVC	<1	5	15	3
PC/ABS	0	6	7	3
PP or PE	0	3	8	2
PC	1	4	1	2
Other	<1	<1	2	<1
Unidentified	5	0	0	3

categories 2, 3 and 4. Thus, WEEE from a CA site will be different from the overall WEEE content.

The actual general materials content of each of these categories can be seen in Table 5. This shows that the plastic-rich streams, in terms of category composition, are categories 2, 3, and 4, which are also the most common in the municipal small mixed WEEE.

There is very little apparent data on the breakdown of the plastic resins used in appliances. The American Plastics Council (APC) conducted two projects[4,5] on characterisation of WEEE plastics. The overall comparisons of the two studies can be seen in Table 6[4] and Table 7.[5]

Table 8 shows a comparison of the specifications of the recycled flame-retardant HIPS from TV casings; the reprocessed T-HIPS by MBA showed similar properties to its virgin counterparts and shows the potential to be used as such.

More detailed breakdowns of the engineering thermoplastics present in products may be obtained by profiling. Computer-product resin mix has a fairly complex composition, and results between this survey and MOEA (see

Table 7 Plastic composition by type, for various WEEE (weight %) (MBA/APC 1999).

Resin	PCs (%)	Fans (%)	Stereos (%)	TV housings (%)	Vacuum cleaners (%)	Other (%)	Total (%)
ABS	34.5	7	5.8	13.7	48		25.8
HIPS	10.2	34.9	27.7	73.3	14.2		19.3
PC/ABS	29.8						16.3
PPO	11.6		19	5.2			8.2
PC	4.5		11.7		11.8	23.5	5.6
PVC	4.9		1.5	3.7	20.1		5.2
PMMA			12.4	1.9		52.9	3.9
PP(TPO)		58.1		2.2		23.5	3.9
SAN	1.8		3.6		5.9		1.8
No ID	2.6		18.2				10.0

Table 8 Comparison of MBA recycled FR HIPS (T-HIPS) properties with those of virgin plastic equivalents.

Resin	Melt flow rate (200/5.0) (g/10 min)	Notched izod impact strength (ftlb/in)	Tensile strength (psi)	Density (g/cm^3)
T-HIPS	7.5	1.5	3100	1.15
Dow Styron 6515	7.5	2.8	2800	1.16
BASF ES 8120	6	2	3500	1.15
Huntsman PS 351	6.5	1.7	4000	1.16

Figure 2) did not agree closely on PC/ABS% in particular. This is also the most rapidly growing market, with an increase from 337 000 tonnes in Europe in 1995 to 595 000 in 2000. The plastic of choice for computer housings is ABS and this correlates to the high content found in WEEE.

9 WEEE Engineering Thermoplastics

The engineering thermoplastics[6] of greatest interest to recyclers are those most prevalent in WEEE streams which retain a high market value.[7–9] These include ABS, PC, ABS/PC, PPO and HIPS. It is useful to know something about these plastics.

9.1 Polycarbonate (PC)

Polycarbonate is a high-quality, engineering plastic with a unique combination of properties including strength, lightness, durability, high transparency and heat resistance, and is easily processed. Hence, it is found in a number of

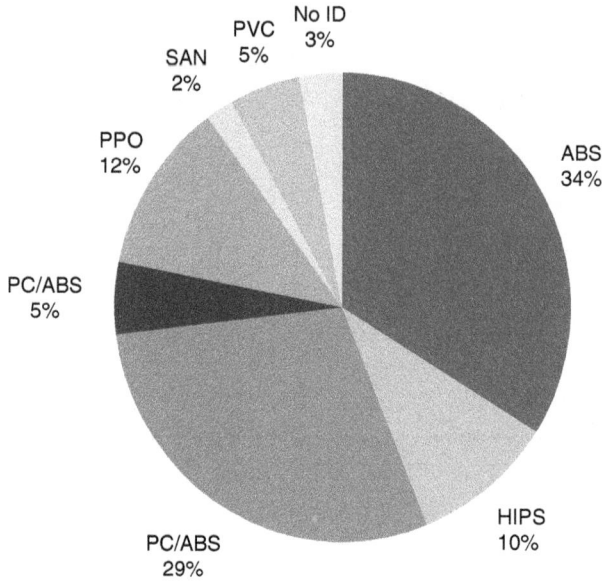

Figure 2   Computer resin composition.

products such as essential medical devices, electronics equipment, computer housings, cars, building and construction applications as well as in consumer goods. It is also extensively used in optical data storage applications (*e.g.* CDs, DVDs), safety equipment and lightweight, transparent roofing in building and construction, power tools and sports goods.

In addition, as polycarbonate is highly durable and lightweight, it is the best choice for applications such as automotive instrument panels, sculpted head-lights and interior trim, which all contribute to light weighting and reduced fuel consumption.

Polycarbonate offers many outstanding characteristics, including[10]

- High transparency, making it ideal for use in protective panelling
- High strength, making it resistant to impact and fracture
- High heat resistance, making it ideal for applications that require sterilisation
- Good dimensional stability, which permits it to retain its shape in a range of conditions
- Good electrical insulation properties
- Biologically inert
- Readily recyclable
- Easy to process

The polycarbonate market is one of the most rapidly growing of the engineering plastics, finding applications in the new technologies with an expected growth rate of 10% p.a. Worldwide production in 2000 had increased to over 1.8 million tonnes from 1.3 Mt in 1998 to 0.6 Mt in 1990. The demand in Europe is in excess of 400 000 tonnes and is valued at around $US 1.6 billion. The details of market size are not readily available for non-commodity plastics but can be obtained from market research reports at a cost.

9.2   ABS (Acrylonitrile-Butadiene-Styrene)

ABS is an ideal material wherever superlative surface quality, colourfastness and lustre are required. ABS is a two-phase polymer blend. A continuous phase of styrene-acrylonitrile copolymer (SAN) gives the materials rigidity, hardness and heat resistance. The toughness of ABS is the result of fine dispersions of polybutadiene rubber particles uniformly distributed in the SAN matrix.

Because of its good balance of properties, toughness/strength/temperature resistance, coupled with its ease of moulding and high-quality surface finish, ABS has a very wide range of applications. These include domestic appliances, telephone handsets, computer and other office equipment housings, lawn mower covers, safety helmets, luggage shells, pipes and fittings. Because of the ability to tailor grades to the property requirements of the application, and the availability of electroplatable grades, ABS is often found as automotive interior and exterior trim components. The market segmentation can be seen in Figure 3.

ABS has the highest production volume of all styrene copolymers with global consumption over 5.4 million tonnes. The growth rate is expected to be above average at around 5.5% p.a. This is expected to take Europe's consumption

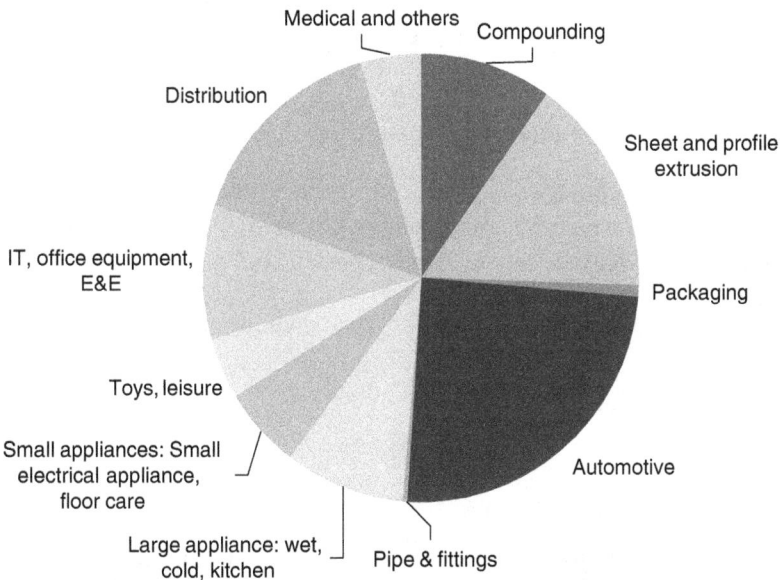

Figure 3   ABS breakdown of consumption into market segment (2004).

from 750 000 to 800 000 tonnes within the next five years. Automotive, appliances and electrical and electronic equipment account for almost 50% of European consumption.

9.3   High Impact Polystyrene (HIPS)

Hardness, rigidity, translucency and high impact strength give HIPS the properties of engineering plastics but at lower cost. It is a graft copolymer of polystyrene with polybutadiene. The rubber modification results in a high-impact polystyrene with a unique combination of characteristics like toughness, gloss, durability and an excellent processability.

Typical applications include yoghurt pots, refrigerator linings, vending cups, bathroom cabinets, toilet seats and tanks, closures, instrument control knobs and consumer electronics.

HIPS figures are not well distinguished from polystyrene. The European market is 2.6 million tonnes a year for polystyrene (2001) of which Europe uses 26% of the global production.

9.4   Polyphenyleneoxide (PPO)

PPO (sometimes referred to as Polyphenylenether – PPE) is a high-temperature-resistant plastic, which is rigid, has good dimensional strength and humidity conditions and is opaque. However, it is fairly brittle and difficult to process and is therefore most commonly found as a blend with HIPS, such as the General Electric Noryl.

The strength, stability and acceptance of flame retardants of PPE (and PPO) make them desirable for machine and appliance housings. The lack of chemical resistance and colour stability means they often have to be painted in these applications. Low water absorption leads to their use in many water-handling products. PPE can also be electroplated for use in automotive wheel covers and grills. Below is a summary of PPE applications:

- Internal appliance components
- Brackets and structural components of office products
- Large computer and printer housings (painted, foamed)
- Automotive wheel covers, plated
- High-tolerance electrical switch boxes and connectors

The market for PPO is not easily distinguished as it is often used in blends with, for example, HIPS.

9.5   PC/ABS Blends

In recent years polycarbonate blends have become increasingly commercially important. PC is widely used in blends due to its excellent compatibility with a range of polymers. Amongst the most significant are those incorporating ABS

(Acrylonitrile Butadiene Styrene). PC/ABS blends exhibit high melt-flow, very high toughness at low temperatures and improved stress-crack resistance compared to PC.

All blends are produced using a compounding step to blend the polymers. This compounding technology is very important for creating the optimal morphology and interaction between the two phases. In combination with the right additive know-how (flame retardant, stabilisation, reinforcement), blends are obtained with an optimally balanced set of properties.

PC/ABS should be used for appearance housings and structural parts which need stiffness, gloss, impact and heat resistance which is higher than ABS, but requiring costs below polycarbonate. Example applications include:

- portable appliances, flashlights, phones
- laptop computer cases
- keyboards, monitors, printer enclosures
- automotive instrument panel retainers
- wheel covers
- small tractor hoods
- non-professional safety helmets

9.6 Flame Retardants in Engineering Thermoplastics

Flame retardants have been in use for over 30 years since it was found that they could effectively reduce the combustibility of most carbon-based materials. Flame retardants have been used in the areas at highest risk of ignition or at greatest fire risk to humans. Thus flame retardants are usually legally required in flammable materials in electronic equipment and furniture. There are over 400 different types of flame retardants and these are generally classified as halogenated (bromine or chlorine) phosphorus-containing, nitrogen-containing and inorganic flame retardants. The brominated flame retardants (BFRs) are currently the largest market group because of their low cost and high performance efficiency. In fact, there are more than 75 different BFRs recognised commercially, of which only 30–40 are widely used.[11]

There are five major classes of BFRs: brominated bisphenols, diphenyl ethers, cyclododecanes, phenols and phthalic acid derivatives. The first three classes represent the highest production volumes. In fact, five BFRs constitute the overwhelming majority of BFR production at this time, although new compounds are being introduced constantly as others are eliminated from commerce. The five major BFRs are tetrabromobisphenol A (TBBPA), hexabromocyclododecane (HBCD) and three commercial mixtures of polybrominated diphenyl ethers (PBDEs), or biphenyl oxides, which are known as decabromodiphenyl ether (DBDE), octabromodiphenyl ether (OBDE) and pentabromodiphenyl ether (pentaBDE). Both penta- and octabromodiphenyl ethers have been phased out in Europe due to concerns over potential toxicity and the formation of dioxins at high temperature, along with the worldwide ban on polybrominated biphenyls in the 1970s. The RoHS Directive also prohibits the two polybrominated diphenyl ethers (PBDEs) pentaPBDE and octaPBDE.

The main concerns that are voiced on brominated flame retardants are:

- evidence of neurotoxic effects – possible reproductive interference
- decomposition of BFRs into more toxic accumulative products
- persistence in the environment
- widespread occurrence of some BFRs being found in animals

A number of BFRs such as decabromodiphenyl ether (DBDE) have been extensively studied for their toxicity and release into the environment. In the case of DBDE formal European risk assessments have found it to be safe.

In the case of TBBPA, the European risk assessment has very recently reported that after an eight-year study of its potential risk to human health and the environment, it is safe to use without restriction in the EU. TBBPA is used in more than 70% of printed circuit boards so this finding is of particular relief to FR4 board manufacturers. Despite this, some local risks were identified at one production plant in Europe where TBBPA is added to ABS and, as a result, the EU approved a Risk Reduction Strategy which recommended an environmental permit to monitor the emissions at this plant. The European Commission's Scientific Committee has also confirmed the EU Risk Assessment conclusions.

Despite these extensive and reassuring studies on DBDE and TBBPA and a number of other leading flame retardants, it should be noted that along with the vast majority of industrial chemicals in use, little detailed toxicity information is available on many of the available BFRs and uncertainty regarding their efficacy has led to their use in European EE manufacturing being curbed with some no longer used. However, they are increasingly being used outside the EU and thus BFRs will be in the WEEE stream for many years to come. The RoHS Directive will mean that any new equipment from 2006 onwards should be free of polybrominated biphenyls and the lower polybrominated diphenylethers.

The European COMBIDENT project found that 22% of samples collected contained halogenated flame retardants, although the validity of the sample collected may not be high or truly representative.[12] The report indicated that the most common halogenated flame retardants, DBDE and TBBA, are likely to be in the concentration range between 5 and 10%, and in most cases are likely to be accompanied by antimony trioxide as a synergist in the range of 3–5%. BFRs are most likely present in HIPS, ABS and ABS/PC components and are more common in small brown goods and IT equipment than large white goods. Note that TBBA is the most common FR used in flame-retarded printed circuit board composites.

A recent analysis undertaken on behalf of WRAP interviewed a number of major manufacturers supplying a range of EE to the UK. This revealed the use of FR as shown in Table 9.[13]

One observation from these data is that streaming of different WEEE product groupings at the input to any dismantling and materials recycling process will lead to very marked differences in the level of FR-containing plastics

Table 9 Use of flame retardants (FR) in various EE products.

Product type	% FR components	FR additive technology	Comments
Printer for PC	10	Brominated	BFR type not banned under RoHS
Telecoms	10	Non-halogen	FR mostly found in mains transformer casing
Power tools	Most parts	Unknown	Non-BFR since 1995
TVs	95	Non-halogen	Polymer blends used to achieve V0 performance
Toys	Not known	Non-bromine	RoHS compliant specified
Computer IT	70–75	Non-PBB/PBDE	Restricted use of brominated FRs since 2002

found. Such an approach could also lead to increased concentration of the common polymer types for each grouping (*e.g.* polypropylene from a power tools stream).

The majority of firms interviewed were at a well-advanced stage of implementing the RoHS requirements. Most large, professional organisations already operate a detailed 'Prohibited Substance List' for all their suppliers of materials and components. It has, therefore, been simple for these companies to incorporate the six chemicals banned under RoHS into these lists. Several firms had already banned brominated flame retardants several years before the start of legislation in response to health and safety concerns voiced in the media. There is some difference in the 'degree of ban' being applied to flame retardants by different firms; this can be demonstrated by the following statements about acceptable materials:

- No halogenated flame retardants to be used in any supplied item
- No brominated flame retardants to be supplied
- BFR types penta-, octaBDE and TBBPA placed on 'Banned' list, but decaBDE and HBCDD on 'Use with Caution' list, if no suitable alternative can be found to deliver required flame retardant performance
- Suppliers simply asked to demonstrate 'RoHS compliance'

10 Materials Composition of WEEE

10.1 Introduction

One of the key challenges in the recycling of end-of-life electronics is the selection of the appropriate recycling technology for the type of product and waste stream being treated. For example, there are clearly very different material compositions in a mobile phone and a washing machine and thus the optimisation of the recycling processes will need to take these differences into account.[14] This section considers three very different specific products (mobile phones, televisions and washing machines) from a materials perspective and highlights their differences and the varying approaches that will be necessary in order to optimise their recycling.[15]

10.2   Mobile Phones

Mobile phones represent what are, in terms of mass and volume, the most valuable electronic products currently found in large numbers in WEEE streams. The rapid introduction of new and improved technology, coupled with increasing functionality such as global positioning systems, wireless, cameras and music players, means that mobile phones have relatively short lifecycles and are often obsolete within a year. In 2005, the UK had more mobile phone subscriptions than people and, as of December 2005, nearly 68 million mobile phones were active in Britain, representing 1.13 active phones for every person. Additionally, there are also estimated to be more than 90 million unused mobile phones in the UK and thus there is a large quantity of valuable materials that will at some point become available for recycling.

When it is not possible to refurbish phones or to re-use individual components, the next level of recycling will involve the recovery of materials. Traditionally, this has tended to mean recovery of valuable metals *via* a smelting process with little regard being paid to the other materials such as plastics, which could be recovered and recycled for use in 'new' polymers.

The polymers used in mobile phone casings are typically materials such as ABS/PC and they are a potentially useful source of polymer that could be re-used. Although it is known that the physical and mechanical properties of polymers degrade both during their service lives and when recycled, the condition of the materials can be determined (see the chapter 'Rapid Assessment of Electronics Enclosure Plastics' by Baird, Herman and Stevens) and the problem can be overcome by blending with virgin material and/or by adding suitable plasticisers and additives.

There have also been processes developed for the separation of materials from end-of-life mobile phones, one such example being the so-called Creasolv process. This process has been developed by the Fraunhofer Institute and CreaCycle GmbH. The Creasolv process is claimed to enable the economical and effective dismantling of mobile phones to give a metal-rich fraction that can be treated using established metal recovery technologies and a polymer solution which can be processed to give a high-quality polymer recyclate. The process uses a tailored organic solvent formulation to dissolve polymers directly from the mobile phone without any mechanical comminution. The solution is then separated from the insoluble materials and the polymer recovered by precipitation.

There is growing interest in the use of non-fossil-fuel-derived plastics for mobile phone cases. Materials such as polylactic acid (PLA), a bioplastic derived from corn, have generated interest as possible replacements for conventional plastics. Until recently, PLA had not been used in electronic devices as it had insufficient heat resistance and strength. However, NEC and UNI-TIKA have jointly developed a PLA bioplastic reinforced with kenaf fibre (kenaf is a biomass-based flexibiliser and reinforcing filler). This new material has been used for the casing of a mobile phone which was launched in Japan during March 2006.

Mobile phones contain a wide range of materials, the specific types and quantities varying from phone to phone. An indication of the materials breakdown for a mobile phone is shown below in Table 10.

It has also been estimated that, in 1 tonne of obsolete mobile phones (excluding batteries), the average non-ferrous metal content is as shown in Table 11.

Mobile phone batteries also contain valuable materials. For example, the cobalt in lithium-ion batteries can be recycled for use in magnetic alloys and the nickel and iron from nickel-metal-hydride and nickel-cadmium batteries can be used in stainless steel. Nickel-cadmium (containing 16–20% nickel) and nickel-metal-hydride batteries (28–35% nickel) used to be the main power sources for phones but lighter lithium-ion batteries (1 to 1.5% nickel) are now favoured.

Between 65 and 80% of the material content of a mobile phone can be recycled and re-used. However, using the plastic as fuel enables the recovery of energy and this can increase the total recovery rate to $\sim 90\%$. Plastics are still considered to be of secondary importance and low value and as yet the infrastructure for recycling polymers, even high-value engineering thermoplastics, has not been fully developed. The metals content of a mobile phone can be

Table 10   Example material content for a mobile phone.

Material	Abbreviation	%
Thermoplastic polymer	ABS-PC	20
Copper	Cu	19
Glass		11
Aluminium	Al	9
Iron	Fe	8
Polymethylmethacrylate	PMMA	6
Silica	SiO_2	5
Thermoset polymer	Epoxy	5
Polycarbonate	PC	4
Silicon	Si	4
Polyoxymethylene	POM	2
Polystyrene	PS	2
Brominated flame retardant	TBBA	2
Nickel	Ni	1
Tin	Sn	1
Liquid crystal polymer	LCP	1

Table 11   Non-ferrous metal content in 1
tonne of obsolete mobile phones.

Metal	Quantity per tonne/kg
Copper	140.00
Silver	3.14
Gold	0.30
Palladium	0.13
Platinum	0.003

successfully recycled. The first stage of recovery is separation of the shredded metals into different fractions. Aluminium, ferrous metal and copper fractions are sold to metal refineries while PCBs are treated in a copper smelter. In the process, copper and precious-metal fractions are smelted and then taken for anode casting and electrolytic refining, which separates copper from the other materials. The remaining precious-metal sludge containing gold, palladium and platinum is sent to precious-metals plants for recovery. Metal recycling has been estimated to save 60 to 90% of the energy required for producing new metal from ore.

10.3   Televisions

Televisions have a unique combination of materials, as well as specific challenges in terms of materials recovery and recycling. They also have long life-times and unique attributes in terms of the materials used in their manufacture. Television technology has also undergone a paradigm shift as manufacturers have switched from cathode ray tube- (CRT)-based units to newer liquid crystal (LC) and plasma flat-panel displays. Thus, from a materials perspective there are many interesting factors that need to be considered in terms of materials selection at the design stage in order to realise the benefits of recycling.

Some of the interesting factors to be considered specifically for televisions and related computer monitors are as follows:

- CRT-based televisions typically have long lives
- CRT-based televisions contain large quantities of glass and hazardous materials (*e.g.* lead and phosphors)
- CRT- and LCD-based television casings contain polymers that could be recycled, if problems with brominated flame retardants can be resolved
- CRT circuit boards are relatively simple, with a low value
- Liquid crystal materials are of very high value
- LCD-based television recycling is currently complex and difficult
- LCD televisions contain other valuable materials, such as indium as ITO

Although nearly all television sales in the UK are now of those with flat panels, televisions typically have long lives and thus there are still large volumes of CRT-based televisions in use. Although the proportion of in-service TVs containing CRTs is predicted to fall to ~52% by 2010 and to virtually zero by 2020, there will nevertheless be CRT-based televisions entering the waste stream for many years to come. In terms of materials composition, plasma and LCD TVs tend to have less glass and more metals and plastics than their CRT equivalents. Modern TVs also feature multiple input and output connectors that are often gold coated, making recovery and recycling more financially attractive. Relative to CRT-based TVs, those with LC displays also have a simplicity and low number of internal components, making them easier to disassemble. The major waste materials issue for this technology is the presence of mercury in the backlights and the liquid crystal materials themselves. A typical LCD unit will contain around 45 mg of mercury, although new back-light technologies are now emerging.

Table 12 below shows the composition of waste materials present in typical televisions of the three main types of technology.

One of the key challenges facing recyclers is the need for tailored treatment technologies for end-of-life flat-panel displays that can accommodate the presence of mercury-containing backlights and unfamiliar and potentially environmentally damaging compositions of liquid crystal mixtures, the exact composition of which remains elusive to recyclers.

Liquid crystal displays are already dominant in the flat-panel display markets, and typical applications also include notebook computers, desktop monitors, electronic organisers, mobile phones, pocket calculators, measuring instruments, electronic toys, digital cameras, audio-video equipment, household appliances and automotive displays. Liquid crystals are thus produced in large quantities and have a predicted growth of ~70% p.a. The total quantity of liquid crystals manufactured worldwide was estimated to be around 600 tonnes in 2005. Although very small amounts are used in each display (typically only about $0.6 \, \text{mg cm}^{-2}$ display area), they can contaminate significant tonnages of aggregated material during disposal. Currently, the only way to deal with WEEE contaminated with liquid crystal displays is by disposing of them in waste incinerators or in landfill. Incineration consumes large amounts of energy and can generate volatile products such as dioxins, while landfilling is known to be hazardous to ground water.

Printed Circuit Boards (PCBs) represent a significant proportion of the overall value obtainable from end-of-life flat-screen TVs. Large concentrations of copper, tin and lead are found in PCBs. Nickel and zinc are also present in significant quantities, although levels of hazardous materials such as arsenic,

Table 12 Waste material content of CRT, LCD and plasma televisions.

	Material content/kg				
	Glass	*Metal*	*Plastic*	*Silicon*	*Total*
CRT	37	4.2	8	4.4	53.6
LCD	3.6	8.4	15	9.6	36.6
Plasma	14.8	12.4	10.9	8.6	46.7

Table 13   Metals in PCB from a flat-screen LCD television.

Amount	Arsenic	Cadmium	Chromium	Copper	Lead	Mercury	Nickel	Zinc
mg/kg	27	9	107	313 000	25 200	0.8	4560	9770

Table 14   Plasticisers and flame retardants found in a PCB from an LCD television.

Material	mg/kg
bis(2-Ethylhexyl)phthalate	607.1
Tetrabromobisphenol A	466

cadmium and mercury are generally very low or below detection limits. Table 13 shows the metal constituents and concentrations in a PCB typically found in a flat-screen LCD television.

It should be noted, however, that the levels of materials such as lead and cadmium should drop, as these were proscribed by the RoHS Directive in 2006. Table 14 shows the levels of plasticisers and brominated flame-retardant material found in PCBs from a flat-screen LCD television.

The use of flame retardants in television cases represents an interesting example of how just one aspect of a product can have a significant impact on the overall environmental credentials of a product and indeed the recycling approaches that can be taken at end of life. Flame retardants are widely used in the polymer casings of CRT-type televisions, but in Europe they were removed several years ago, despite finding continuing use in TVs made in other non-European countries. The removal of flame retardants in TV casings led to a potential environmental benefit in terms of avoiding the emission of persistent brominated species to the environment but it has also led to an increase in the number of fires caused by faulty TVs. It has been estimated that at least 16 people, and perhaps as many as 160 people, die each year in Europe as a result of house fires starting in televisions and this trend will grow if fire protection is not assured. Without flame retardants, the polymers represent less of an environmental hazard and they are also easier to recycle and have higher value. However, when this is traded against consumer safety and possible loss of life it is necessary for society to decide what balance point is acceptable.[13]

Another materials-related challenge with the recycling of conventional CRT-based televisions relates to the glass that is used in the picture tube. This glass is essentially of two compositions, with the funnel glass containing lead while the front-panel glass is lead-free. There are large quantities of this glass appearing as increasing numbers of CRT televisions reach end of life and there are questions about the best routes for recycling the glass and subsequent re-use opportunities, especially as closed-loop opportunities are diminishing with the demise of CRT manufacturing. It is also interesting to note that only two material sources in a CRT-based television have a positive net value after dismantling at end of life and these are the copper-bearing yoke and the electronics. Overall, it seems that the only truly cost-effective approach to disposing of CRT-based televisions is *via* landfill, but this will no longer be possible in the traditional sense, because of the introduction of legislation such as the WEEE Directive.

10.4 Washing Machines

Washing machines represent the opposite end of the WEEE spectrum in terms of size, mass, material composition and value relative to mobile phones and TVs. They are regarded as typical white goods and traditional recycling activities have been focused on metals recovery. Washing machines, even today, contain only a relatively small amount of electronics and processing power and, unlike the high-value circuit boards found in mobile phones, these are typically made from low-cost laminates containing fewer valuable materials. The materials composition of a typical Japanese washing machine is given in Table 15.

It should be noted that European washing machines typically contain a concrete stabilising weight whereas Japanese machines use a plastic container filled with salt water. Most of the environmental impact of a washing machine occurs during the use phase. Solid waste is generated at a number of stages, such as when the original packaging is removed and disposed of and also at end of life when disposal takes place. However, while the solid waste levels at these stages are significant, they were found to total less than 15% of the total solid waste produced by the washing machine. This is because of the large number of washing powder packets and other aids that are consumed and disposed of during the machine's life. It is also interesting to note how small the contribution is at end of life to the overall impact of the washing machine. Clearly, significant opportunities exist at the use phase, such as the use of lower washing temperatures and the development of new detergents that require less material to achieve a given function. During the use phase the washing machine consumes large quantities of water and produces polluted water. Consequently, there are also opportunities for water consumption reductions during use that would have a significant impact on the overall environmental impact.

Overall, then, it is clear from a whole lifecycle perspective that the largest environmental impacts of a washing machine come from the use phase, largely because of the quantities of electricity, water and detergents, *etc.* that are

Table 15   Materials composition of a typical Japanese washing machine

Material	Content/wt%
Steel	53
Plastic	36
Copper	4
Aluminium	3
Other materials	4

needed. Recycling, of what is effectively more than 50% by weight of steel, is essentially about metals recovery. In addition to the large amount of steel components, information from Panasonic's METEC facility indicates that there are three additional material streams from end-of-life washing machines: plastics, mixed other metals and a mix of materials that are often simply incinerated. Opportunities exist to do more with the plastics components.

11   Conclusions

We began this chapter by highlighting the need to consider the many factors affecting materials choice in the context of whole-life management. We could extend this to include extended multiple lifecycle use as it is clear that there are many materials streams that require extended lifecycle management. This is perhaps best exemplified by the need to recover and re-use the non-ferrous metals and particularly those with increasing pressure on conservation, such as the rarer platinum metals and ITO. With dramatically increasing commodity prices, the same need exists for economic reasons with the more common engineering thermoplastics, where opportunities for closed-loop or in-sector recycling exist.

Primary environmental and sustainability-centred legislation arising in different global regions of the world and the economics of materials supply and management across the whole lifecycle are the primary materials drivers. The selection of materials and additives that can continue to be used sustainably in the future, and those which need to be developed to replace existing materials that cannot meet the requirements, is a global challenge facing the electrical and electronics industry and their supply chains.

Such selection cannot be based solely on the requirements of current legislation as this is certain to change in the future. It must be based on sound principles of sustainable production and use which must satisfy current and future technical, economic, environmental and social criteria. As discussed, there may be no net benefit to society if a perceived environmental benefit is obtained with no or only a small risk reduction to human health and environment, at the expense of an increased risk of loss of life because of reduced fire performance.

Such thinking has to examine how a progressive move to a more sustainable materials future can be achieved, while accepting that historic legacies must also be managed.

Current legislation and public pressure are providing a force for change in the right direction and producer responsibility directives such as WEEE, coupled with the RoHs, and EuP directives in addition to REACH, are facilitating this change. But care is required to be balanced in the proscription of materials and additives and an appropriate balancing of consumer protection and environmental protection has to be achieved. This can be achieved by both regulated and unregulated responsible materials stewardship and there is a place for industry-promoted voluntary initiatives working alongside prescriptive legislation.

So REACH regulations will probably lead to the demise of some important chemicals, requiring the formulation of many new materials used in electrical and electronic applications. However, these new materials must have clear and demonstrable benefits over existing materials and not compromise other very necessary technical, economic and human-safety performance requirements. In contrast, the Energy using Products (EuP) Directive will require the electronics industry to take a more holistic approach to the way it manufactures its products, with emphasis being placed on all aspects of a product's lifecycle from eco-design to end of life. With its focus on energy consumption, this is unlikely to present conflicts to human health and safety.

The encouragement of eco-design principles, which include materials selection, will lead to the integration of environmental considerations during the design and materials-selection phases of a product. While actual measures have been decided for specific products under the supervision of a designated panel of EU member state experts, it is important that the principles are extended to both products and materials in general. This will require changes in thinking and practice within the electronic and recycling industries which pick up the challenge at end of product-life. Some of the issues that industry will need to come to terms with are:

- managing materials restrictions and limitations
- improving end-of-life recovery and re-use, and increasingly move to selective materials recycling
- improving resource and energy efficiency of products in manufacturing and use
- adoption of a more holistic approach to product lifecycle from design through to end-of-life management
- development of more sustainable materials sourcing and management practices, with improved materials stewardship along the supply chain
- the development of new materials that confer net economic, environmental and social benefits with no compromise of human safety

Ideally, the industry and regulators will achieve some degree of consensus on how to address these major challenges.

References

1. T. Muratani, and T. Honma, Treatment apparatus for waste panel and method for treating waste panel, WO 2006/115105 A1, 2006 (SHARP KABUSHIKI KAISHA).
2. Status report on waste electrical and electronic equipment in the UK 2005, ICER, 2005.
3. Trial to establish waste electrical and electronic equipment (WEEE) protocols, DEFRA/CIWM, 2007.
4. 4 M. M. Fisher, M. B. Biddle, T. Hainault and D. S. Smith, *Characterization and Processing of Plastics from Minnesota's Demonstration Project*

74 *Gary C. Stevens and Martin Goosey*

for the Recovery of End-of-Life Electronics, American Plastics Council, Sony Electronics Inc., David J. Cauchi, Consultant, David A. Thompson, 2000.

5. *Recovery of Plastics from Municipally Collected Electrical and Electronics Goods*, American Plastics Council, 1999.
6. British plastics federation (www.bpf.co.uk).
7. M. Fisher, J. Biancaniello, T. Kingsbury and L. Headley, *Ten Facts to Know about Plastics from Consumer Electronics,* American Plastics Council (APC), 2000.
8. *The Characteristics of Plastics-rich Waste Streams from End-of-life Electrical and Electronic Equipment*, Plastics Europe, 2006.
9. *Plastics. A Material of Innovation for the Electrical and Electronic Industry*, Plastics Europe, 2001.
10. Plastics Europe (http://www.plasticseurope.org/Content/Default.asp?PageID = 1188).
11. G. C. Stevens and A. H. Mann, *Risks and Benefits in the Use of Flame Retardants in Consumer Products*, DTI report URN 98/1026; produced by the Polymer Research Centre, University of Surrey, 1999.
12. *High Quality Plastic Materials from Electronic Waste by Use of Combined Identification Methods and New Handling Technologies (COMBIDENT)*, BRPR-CT98-0778, Fh-ICT, 2001.
13. *Develop a Process to Separate Brominated Flame Retardants from WEEE Polymers*, WRAP, 2006.
14. *WEEE Flows in London report*, London remade/Axion, 2006.
15. R. Kellner, *An Integrated Approach to Electronic Waste (WEEE) Recycling*, KTN/DEFRA, 2006.

Dumping, Burning and Landfill

IAN HOLMES

1 Introduction

This chapter will look at some of the historic methods of waste disposal. Therefore the descriptions of these disposal methods may appear fairly general.

Until relatively recently there was little distinction between electrical waste and any other form of municipal waste. Even today, a significant amount of small electrical waste is disposed of in household wheelie bins and usually finds its way to landfill or incineration with the other municipal solid waste (MSW). Despite the current drives to improve recycling rates, landfill and incineration are still the main disposal routes in the UK. In recent years there has been a host of regulations and public reactions affecting these disposal routes, and these have been some of the main drivers for the development of new technologies to enable the recovery and recycling of much of the waste streams. It now seems that these efforts are paying off. We can see from the charts below (see Figures 1–3) that waste to landfill is beginning to decrease and recycling rates are rising. These charts are taken from the Environment Agency Waste Data pages.

From Figure 1 it can be seen that:

- landfill inputs are down by 15% over the whole period
- incineration inputs increased 52% (1.3 million tonnes) as additional capacity came on stream in cement kilns and other new facilities. Overall hazardous waste capacity increased by three times and energy recovery capacity doubled
- there was an apparent doubling of treatment inputs, with an increase of 14 million tonnes. This is, however, somewhat misleading, as 6 million tonnes of this increase occurred at one waste/water treatment facility in the

Issues in Environmental Science and Technology, 27
Electronic Waste Management
Edited by R.E. Hester and R.M. Harrison
© Royal Society of Chemistry 2009
Published by the Royal Society of Chemistry, www.rsc.org

Waste management trends by facility type 2000/1 to 2005

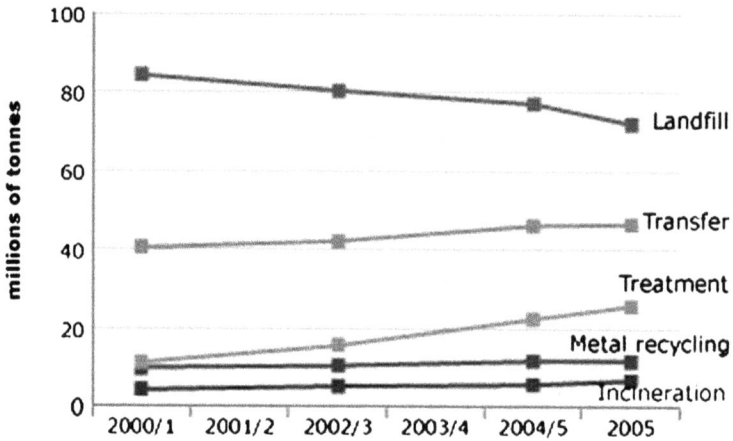

Figure 1   Waste management trends in the UK.

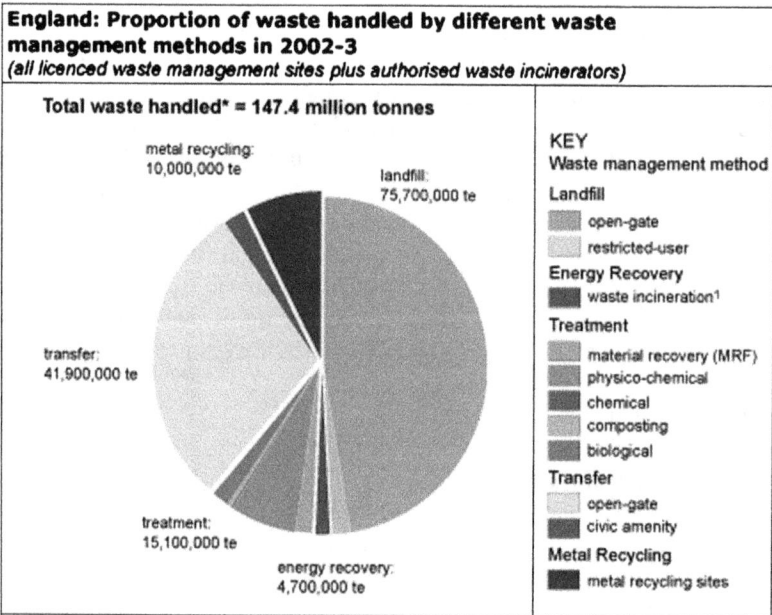

Figure 2   Proportion of waste management methods by type 2002–2003.

North-East. A better estimate of treatment growth over the period would be around 60%.[1]

• 72 million tonnes of waste was deposited at landfill sites in England and Wales in 2005. This compares to 75 million tonnes in 2004/5 (last reporting period) – a reduction of around 5%. Overall, landfill deposits have fallen by 15% since 2000/1.

England & Wales: Proportion of waste handled by different waste management methods in 2005
(all licenced waste management sites plus authorised waste incinerators)

total waste handled[4] = 166 million tonnes
treated or disposed of[3] = 119 million tonnes
deposits at transfer sites = 47 million tonnes

transfer:
total handled
= 46.5 m te

landfill: total deposits
= 71.9 m te

metal recovery:
total handled
= 11.7 m te

treatment:
total deposits
= 25.7 m te

lagoon/borehole:
deposits = 2.9 m te

waste incineration:
total deposits = 6.8 m te

KEY
management method

Landfill
  merchant
  restricted-user
  lagoon/ borehole[3]

Waste incineration[1]
  Incineration

Treatment
  MRF
  Physico-chemical
  Chemical
  Composting
  Biological

Metal Recovery
  Metal recycling

Transfer
  Open-gate
  Civic amenity

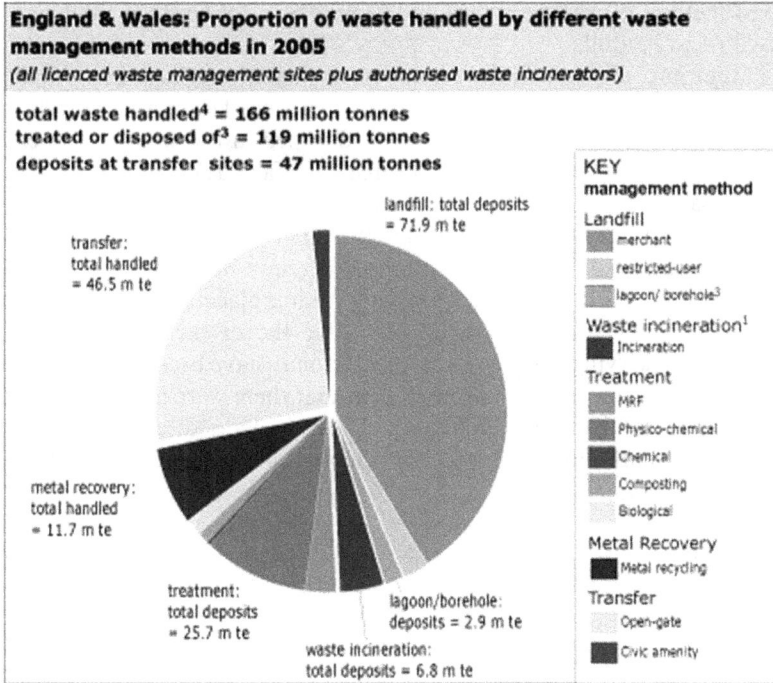

Figure 3 Proportion of waste management methods by type 2005.

1.1 England: Site Inputs 2002–2003

Almost half of all the waste handled at licensed or authorised waste management facilities was landfilled. Less than 5 million tonnes was incinerated. A total of 15 million tonnes was treated. Almost 42 million tonnes, 30% of all waste handled, was transferred before being managed and disposed of elsewhere. See Figure 2.

1.2 Waste Inputs to Different Management Options in 2005

During 2005, 28% of the total inputs to permitted waste management facilities went to a transfer facility, followed by treatment or disposal. Inputs to waste transfer should be excluded from waste production estimates to avoid double counting. See Figure 3.

Among the treatment and disposal options 60% went to landfill, 22% for treatment, 6% for incineration and 10% went into metal-recycling facilities.

2 Landfill

2.1 Historical

Current thinking is that landfilling of waste is the last resort for waste disposal; however, this has not always been the case. In spite of this attitude, it is still the cheapest method of waste disposal and around 80% of waste in the UK is still disposed of to landfill.

The 'dumping' of waste in the ground goes back many years in the UK. The industrial revolution and rapid urbanisation of areas led to a large increase in domestic and industrial waste often being dumped in the streets. This created risk to human health from vermin and flies. As far back as the 12th century, laws were passed to control the dumping of waste and to ensure it was moved to specified 'dumps' areas away from inhabited areas.

The most basic landfill was placing the waste in a hole in the ground. With the large amount of mining and quarrying taking place in the UK, much use has been made of the empty holes left after the extraction of coal, other minerals and quarrying activities. Initially this may have been seen as a wise use of resource; however, it also became evident that there were many downsides to this approach to disposal (Figure 4). These include disease, health risks, environmental contamination and the danger of explosion from untapped landfill gas (methane).

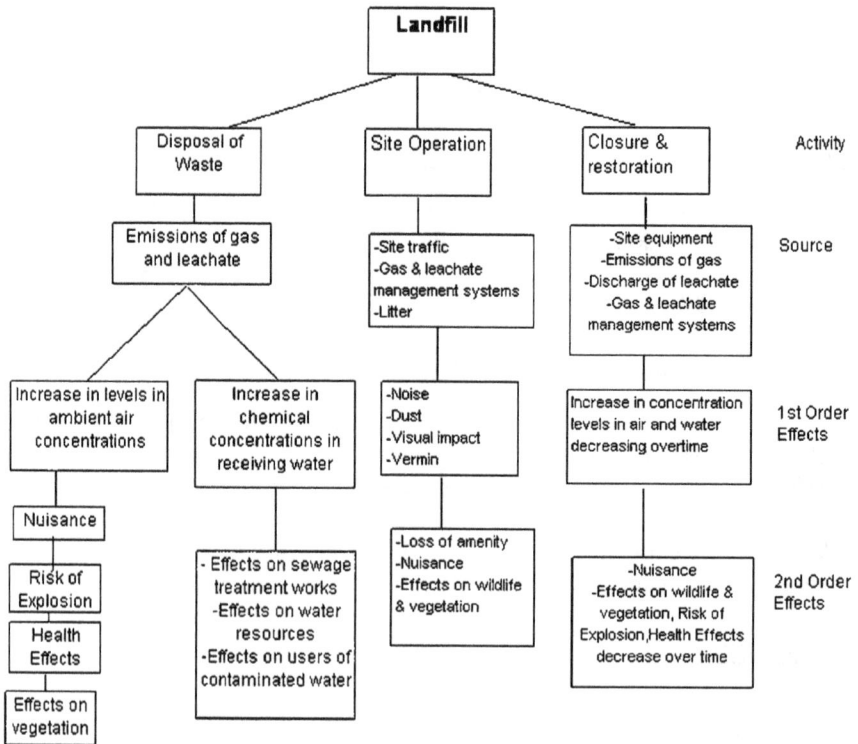

Figure 4   Impacts and effects of landfill on the environment.

2.2 Pollution from Landfills

There are two main pollutants produced by the landfilling process, both of which can have serious effects on the environment. These are landfill gas and leachate.

2.3 Landfill Gas

Landfill gas can contain up to 55% methane (CH_4) and 45% carbon dioxide (CO_2). The problems associated with methane production have previously focused on the flammability of the gas and looked to collect the gas in order to prevent explosions such as the Loscoe Explosion, Derbyshire.

During March 1986 an explosion occurred in a bungalow near to the landfill site at Loscoe. The explosion destroyed one bungalow, and the resulting investigation showed two more houses had been unfit for habitation due to high levels of methane for several months.[2]

However, with the current information on climate change and greenhouse gas emissions, it is now even more important to control and collect landfill gas. Although CO_2 is used as the main indicator in discussions around carbon footprints, studies have shown that CH_4 has between 20 and 60 times the Global Warming Potential of CO_2.[3] Luckily, CH_4 has significant value as a fuel and is usually used on-site to produce energy or make biogas fuel.[4] As a last resort it can be flared off to produce CO_2 and H_2O.

2.4 Leachate

The leachate is produced as water percolates through the waste within a landfill site. The water extracts soluble chemicals and products of decomposition. The two main components are organic compounds and heavy metals. The presence of organic chemicals is caused by the breakdown of organic waste and also directly from discarded organic chemicals in the waste, such as pesticides and alcohols. This alters the pH of the water, making it acidic, which in turn allows metal ions, which are insoluble at neutral pH, to dissolve in the liquid.

Many of these metals are from the discarded electrical equipment within the waste. This can lead to mercury, lead, chromium, copper, zinc and cadmium being present in landfill leachate. Despite the WEEE Directive and the forthcoming Batteries Directive, batteries and small electrical or household items are still likely to end up in landfill in the UK.

Both of these contaminants are also produced in waste 'dumps' such as the ones in Lagos, Nigeria. Lagos is a popular destination for some e-waste as it can be disassembled and recycled by cheap labour. According to the Basel Action Network (BAN) environmental group, around 500 shipping containers, with a volume equivalent to 400 000 computer monitors, enter Lagos each month.[5] These shipments are stockpiled in open dumps around the city, awaiting disassembly, where the conditions allow the production of toxic leachate which contaminates the local water supplies and soils. (Figure 5a and b)

(a)   (b)

Figure 5   (a) Piles of circuit boards from hazardous computer waste stretch into the distance near an e-waste scrap yard. The circuit boards will be smelted by hand to extract metals. Smelting releases highly poisonous gases and pollutes the environment. © Greenpeace. (b) Mountains of e-waste. The owner of an e-waste scrapping yard stands in front of a mountainous pile of computer waste waiting to be scrapped to recover useful plastics and metals. © Greenpeace/Natalie Behing.

Early landfills also had uncontrolled release of contaminants into the environment, which led to problems of land and water contamination in surrounding areas. The EC Landfill Directive[6] introduced in 1999 aims 'to prevent or reduce, as far as possible, negative effects on the environment, in particular, the pollution of surface water, groundwater, soil and air, and on the global environment, including the greenhouse effect, as well as any resulting risk to human health, from the landfilling of waste, during the whole lifecycle of the landfill'. This directive encourages the reduction of waste going to landfill and also stipulates a system for issuing permits for the operation of landfill sites and their design.

2.5   Landfill-site Construction

The two most common types found in the UK are controlled-contaminant release (disperse and attenuate) and containment and collection of contaminant.

2.5.1   Attenuate and Disperse

This type of landfill is no longer constructed as it is not acceptable under current legislation and its use is decreasing.[7] These landfills are unlined and there is an uncontrolled release of contaminants (Figure 6). The dispersal of the leachate is influenced by the geology of the land and these allow a slow movement of the leachate away from the site over a long period of time. This dilutes and reduces the toxicity of the fluids.

However, this type of landfill design is not suitable to all types of geographical location and, to prevent leachate contaminating groundwater supplies, careful surveys must be conducted.

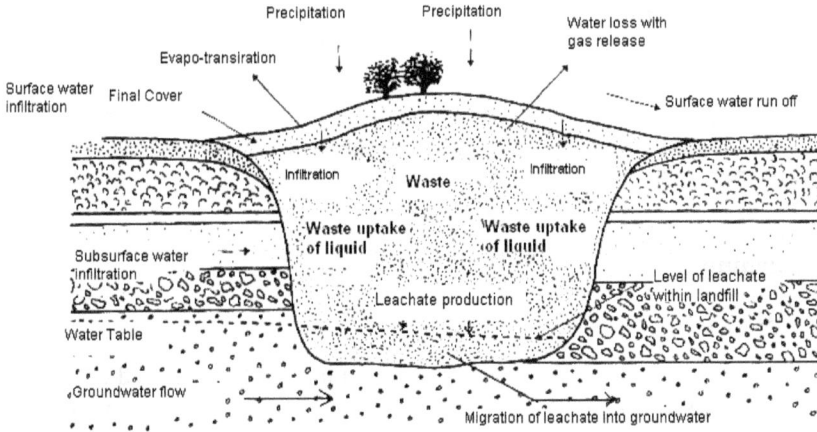

Figure 6 Schematic diagram of the water balance for an attenuate and disperse landfill site. Source: WMP 26, Landfilling Wastes, Dept. of the Environment HMSO.

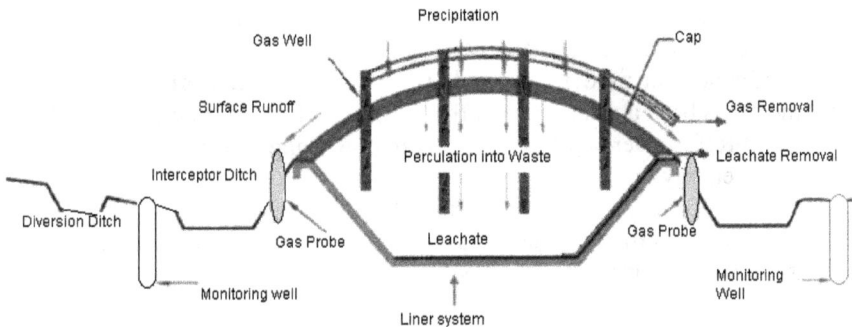

Figure 7 A schematic of a typical containment landfill with leachate and gas collection.

2.5.2 Containment Landfills

Modern landfills are strictly regulated and technically designed to minimise any detrimental effects of the site. These landfill regulations demand strict requirements for all stages of a landfill site from location, design and operation through to 'closure' at end of life of the sites, as well as the control of emissions such as gas and leachate.

The majority of UK landfills are now containment landfills (Figure 7). They are lined with a material that contains the leachate within the landfill-site boundary. The degradation processes take place within the landfill until stabilisation is complete. The leachate and landfill gas are able to be collected and treated.

In some landfill designs the leachate produced is recirculated through the waste where it acts as an accelerant in the degradation process. This is known as a flushing bioreactor.

In other types of site the leachate is removed for treatment either on or off site, or may even be discharged into the sewerage system where the correct permits are in place.

3   Burning

3.1   Historical

Like dumping in a hole, waste disposal by burning has been carried out in some form or other since man discovered fire. Burning waste reduces the volume and converts the organics content, which would decay and smell, into ash. However, if the waste was not sorted or segregated prior to incineration, the outputs from the combustion process were often toxic stack emissions and ash containing heavy-metal residues.

The early waste incinerators were designed just to burn the waste, reducing the volume and producing an inert ash which could be sent to landfill. The first recorded one in the UK was in Nottingham in 1874.[8] Although as far back as 1885 incinerators were being used to generate power, the majority were built purely to reduce the volume of waste.

With advances in modern technology and new thermal-processing techniques, there has been a steady rise in the 'burning' of waste. Energy from Waste (EFW) and Refuse Derived Fuel (RDF) are now being considered as sources of 'renewable' energy.

As shown in Figure 1 earlier, incineration increased by 52% (1.3 million tonnes) between 2001 and 2005, as additional capacity came on stream in cement kilns and other new facilities came on line.

3.2   Incineration

This is the most basic form of thermal treatment of waste. It can be carried out with or without energy recovery. Older incineration plants tended to use Mass Burn technology. This consisted of the delivery of unsorted waste into a storage bunker where it could be fed into the combustion chamber. These early municipal incinerators rarely included energy recovery of any sort. For example, in 1991 there were around 30 large-scale mass-burn incinerators in the UK but only 6 of them were equipped with any form of energy recovery. These older plants also had limited pollution-control systems which led to the closure of a number of them when the EC introduced two Directives on the Reduction (for existing plant[9]) and Prevention (for new plant) of Air Pollution from Incinerators.[10]

3.3   Mass Burn

This is the most common form of Municipal-Waste Incinerator (Figure 8). It is simple and relatively easy to operate. Waste is not pre-sorted; deliveries deposit

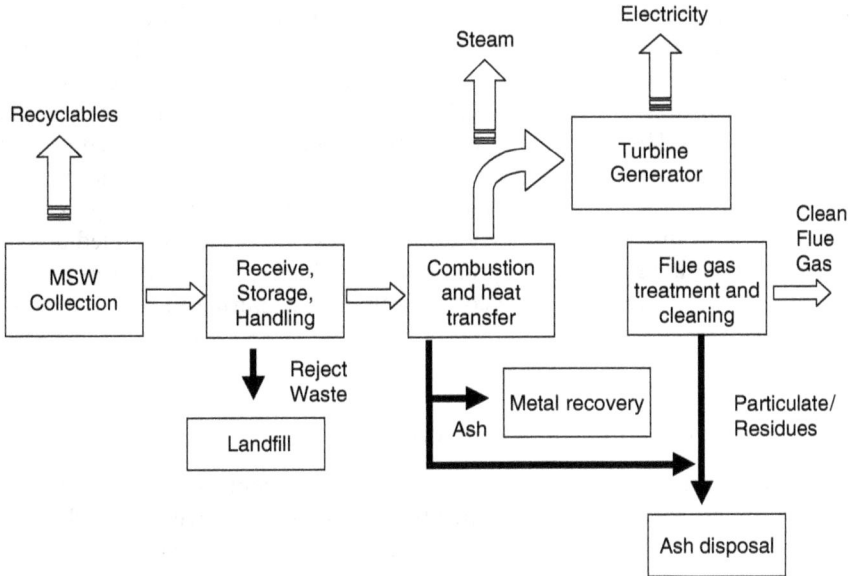

Figure 8 Block diagram of a mass-burn incinerator.

the waste in a storage hopper. It is fed from the hopper onto a moving grate that moves the waste down through the combustion chamber. The waste is agitated so that it is fully burnt and then discharged into the ash pit at the end of the process. If energy-recovery systems are fitted, then it is possible that the combustion of one tonne of waste could produce between 550 and 650 kilowatt hours of electricity.

The mass-burn incinerator (Fig. 8) involves the following stages:

- Mixed waste is collected from source and recyclables may be removed.
- It is then dumped into a holding area.
- The waste is fed into the incinerator.
- The combustion of the waste heats a boiler and the steam generated drives a turbine to produce electricity. Waste heat and steam may be recovered for other applications.
- The heaviest ash passes through a grate where it has some metal-content extracted by electromagnet.
- Flue gases and fine ash are scrubbed in a reactor to remove SO_x and NO_x and dioxins.
- The final stage is a fine-particulate filter system to remove fine particulates; then the cleaned gases are released into the atmosphere through the stack.

3.4 Energy Recovery/Energy from Waste (EFW)

In plants which are designed specifically to recover energy from the combustion of waste, Fluidised-bed Combustion may be used. The reactor design is based on the similar technology used in coal-fired power stations. This method pre-processes the waste to remove the non-combustible and recyclable materials. The sorted waste is then shredded to produce a coarse Refuse-derived Fuel (RDF). This processing gives the RDF a higher calorific value than that of the untreated waste. The combustion bed of the incinerator is constructed of inert material such as sand or ash. The combustion air is blown through this bed at a rate sufficient to create a rapidly moving or 'fluidised' bed. This design greatly improves the combustion efficiency of the process. This has the added advantages of generating more energy and reducing pollution. However, this process is up to 35%[11] slower and more complex than mass burn and therefore more expensive.

3.5   Advanced Thermal Processing

Relatively recent developments in the thermal treatment of waste involve the heating of the waste to high temperatures in conditions with limited or no oxygen present. This atmosphere causes the waste to break down into a carbon-rich char, oil and a synthetic gas (syngas). There are two main processes: gasification and pyrolysis. They are similar but have some slight differences in the way the waste is treated within the process.

3.5.1   Gasification

The gasification process is actually not a new concept. It was used to produce 'town gas', which was the main source of heating and lighting gas in the UK during the 1940s and 1950s, prior to the development and distribution of natural-gas supplies.

Gasification involves the reaction of the waste at a high temperature with a controlled amount of oxygen. The carbon content of the waste reacts with the oxygen to produce carbon monoxide and hydrogen. This resulting gas is known as syngas and is used as a fuel. The use of syngas as a fuel is far more efficient than direct combustion of the waste in a mass-burn incinerator.

The syngas can be used to generate electricity *via* a gas turbine, which is less expensive and more efficient than the steam cycle used in incinerators. Due to the reduced volume of oxygen used in the process, less gas is produced than flue gas, so a smaller volume needs cleaning as compared with incineration.

The gasification process is self-sustaining and requires some air or oxygen for the partial combustion of the waste.

The gasification reactions are as follows:

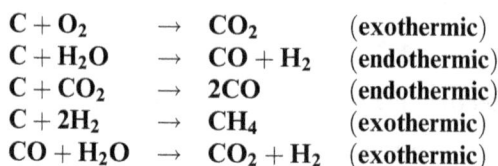

$$
\begin{array}{lll}
C + O_2 & \rightarrow & CO_2 & \text{(exothermic)} \\
C + H_2O & \rightarrow & CO + H_2 & \text{(endothermic)} \\
C + CO_2 & \rightarrow & 2CO & \text{(endothermic)} \\
C + 2H_2 & \rightarrow & CH_4 & \text{(exothermic)} \\
CO + H_2O & \rightarrow & CO_2 + H_2 & \text{(exothermic)}
\end{array}
$$

3.5.2   Pyrolysis

Pyrolysis is the thermal processing of the waste in the complete absence of air. It requires an external heat source to drive the endothermic pyrolysis reactions. This process is also known as destructive distillation.

The three main outputs of the process are:

- A gas – contains mostly CH_4, CO, CO_2 and H_2.
- A liquid fraction – an oil/tar stream containing acetic acid, acetone, methanol and complex oxygenated hydrocarbons.
- A char – almost pure carbon, plus the inert content of the waste.

These three fractions are all usable as a source of fuel for other combustion processes.

These processes are still being developed and fine-tuned to become economically viable for waste processing. However, the recent increases in landfill tax and the diminishing availability of landfill in some regions have begun to make them attractive to local authorities. Some recent examples follow.

3.5.3   Compact Power

The Compact Power process uses a combination of gasification and pyrolysis and high-temperature oxidation to convert the waste into fuels and other usable products. The process is designed to produce energy from waste and also saleable fuel products to industry. It was the first commercial plant to receive an Integrated Pollution Prevention and Control (IPPC) licence from the Environment Agency, at Avonmouth near Bristol. However, Compact Power has recently gone into administration and its assets and intellectual property were acquired by Ethos Recycling.[12]

3.5.4   Graveson Energy Management Ltd (GEM)

GEM has patented an advanced thermal conversion process which is described as 'flash pyrolysis'. It uses the process developed in the petrochemical industry known as 'cracking'. This is the 'destructive distillation' or chemical decomposition of feedstock by intensive heat in the absence of oxygen atmosphere. The company was originally based in Hampshire but has recently relocated (April 2007) to Port Talbot, South Wales.[13] The technology is available as modules which can be incorporated within power-generation systems. GEM have secured several contracts to supply this technology for new energy-from-waste plants around the UK. These include dedicated food- and packaging-waste plants as well as municipal solid-waste (MSW) treatment plants.

3.6   Pollution from Incineration

There are a number of substances produced by the incineration process, several of which have a direct effect on human health, others an indirect effect by

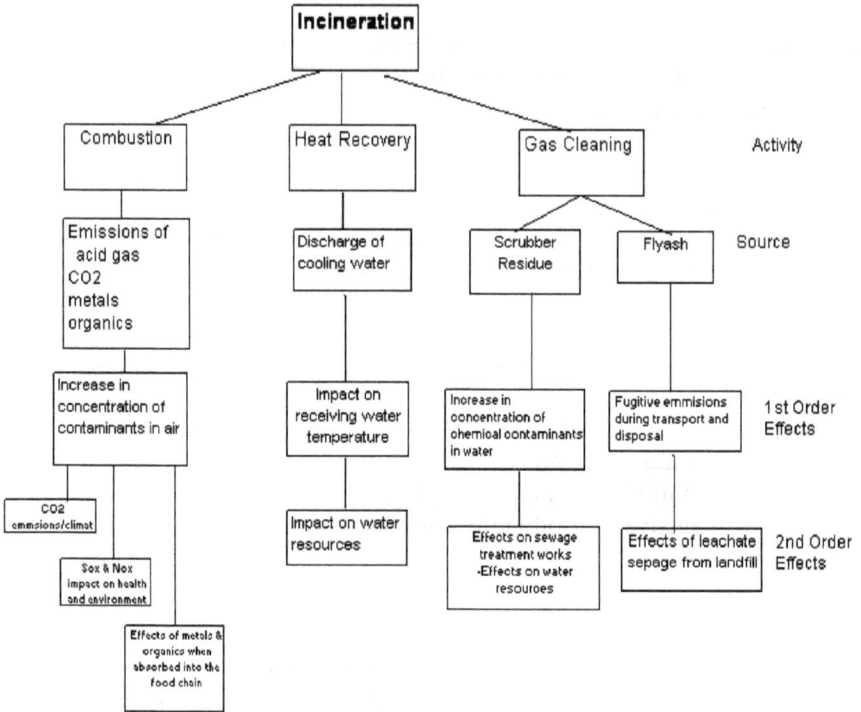

Figure 9   Impacts and effects of incineration on the environment.

damaging the local and global environment (Figure 9). Some of the more unpleasant substances can be directly related to the combustion of WEEE, such as dioxins (PVC combustion), heavy metals (lead and mercury from circuit boards and LCDs), fluorinated compounds (LCDs and plastics).

Dioxins.   The term 'dioxin' is used to refer to the polychlorinated dibenzo-p-dioxins (PCDD), and 'furan' is used for the polychlorinated dibenzofurans (PCDF). These chemicals are defined as tricyclic, aromatic compounds formed by two benzene rings connected by two oxygen atoms in polychlorinated dibenzo-p-dioxins and by one oxygen atom and one carbon-carbon bond in polychlorinated dibenzofurans, the hydrogen atoms of which may be replaced by up to eight chlorine atoms (UNEP, 2001).

Dioxins are highly persistent trace chemicals found in soils, sediment in freshwater and the sea, plants and animals – including humans. They belong to the family of chemicals known as 'Persistent Organic Pollutants' (POPs), which are subject to 'The Stockholm Convention',[14] an international treaty to reduce exposure to POPs which was ratified in May 2004. Key characteristics are their resistance to decomposition (persistence) in the environment, their ability to become more concentrated through the food chain (bioaccumulation) and toxicity.

Heavy Metals. Heavy metals are characterised as chemical elements with a specific gravity that is at least five times the specific gravity of water. There are 23 'heavy metals': antimony, arsenic, bismuth, cadmium, cerium, chromium, cobalt, copper, gallium, gold, iron, lead, manganese, mercury, nickel, platinum, silver, tellurium, thallium, tin, uranium, vanadium and zinc. Many of these are commonly found in waste electrical equipment and will therefore be emitted in combustion gases or found in the waste ash from incineration. These metals can affect human health in a variety of ways. Small amounts of these elements are found in our diet and some are actually necessary for good health, *e.g.* zinc, iron, manganese; however, large amounts of any of them can cause poisoning. The toxic effects of heavy metals in the human body can result in malfunctioning of the mental and central nervous-system function, low energy levels and damage to blood composition, lungs, kidneys, liver and other vital organs. Long-term exposure to some elements has been seen to result in slowly progressing physical, muscular and neurological degenerative processes that have similar symptoms to Alzheimer's disease, Parkinson's disease, muscular dystrophy and multiple sclerosis. Contact with heavy metals can also exacerbate allergies and long-term contact may even cause cancer.

Particulates. Particulates are tiny particles suspended in the air. They consist of a wide range of materials. As they vary in size and shape they are often split and reported as total particulates or PM10 (particles with diameter less than or equal to 10 micrometres). PM10 are generated from three main sources: primary particles from combustion sources (incineration); secondary particles created by atmospheric chemical reactions; and suspended soils, dusts, sea salt and biological particles which form coarse particles.

There have been a number of studies on particulate pollution which have linked it with acute changes in lung function and respiratory illness,[15] resulting in increased hospital admissions for respiratory disease and heart disease, absences from respiratory infections or aggravation of chronic conditions such as asthma and bronchitis.[16]

Inorganic Acidic Gases. These can include hydrogen chloride (HCl), hydrogen fluoride (HF), hydrogen bromide (HBr), sulfur oxides (SOx) and nitrogen oxides (NOx). The gases form corrosive acids on contact with water in the atmosphere or on surfaces. Acid gases can have a direct effect on health in large concentrations, mainly related to respiratory complaints such as asthma and breathing difficulties. These acids can cause damage to the surfaces they come into contact with, be they metal, stone or organic; they will also alter the pH level of water courses. The increasing content of plastic in waste has led to higher levels of HCl and HF.[17]

Carbon Dioxide. The levels of CO_2 produced do not have direct health implications. However, CO_2 is a greenhouse gas and is known to contribute to climate change and global warming.

4 Legislation Summary

Over the last ten years or so there has been a raft of legislation affecting the disposal of waste to landfill and regulating the use and emissions of incinerators. This has been driven by a number of factors:

- The need to protect the environment from pollution
- The need to reduce the emission of 'greenhouse gases'
- Safeguarding human health
- Increasing the volumes of recycling of valuable resources

This legislation has been driven by EC directives, international conventions and treaties, and implemented by the UK Government to suit its requirements.

4.1 Current UK Legislation

The legislation of waste disposal is extensive and has become more so in recent years. Although much of this legislation targets general waste disposal there are also several pieces which are specifically aimed at the disposal of products containing electrical components.

This list with active links to all the relevant legislation is available on the Environment Agency NetRegs website Waste Legislation page (http://www.netregs.gov.uk/netregs/legislation).

Control of Pollution (Amendment) Act 1989
Environment Act 1995
Environmental Protection Act 1990
Controlled Waste Regulations 1992 SI 588
Controlled Waste (Amendment) Regulations 1993 SI 566
End-of-Life Vehicles Regulations 2003 SI 2635
End-of-Life Vehicles (Producer Responsibility) Regulations 2005 SI 263
Environmental Protection (Duty of Care) Regulations 1991 SI 2839
Environmental Protection (Duty of Care) (England) (Amendment) Regulations 2003 SI 63
EU Regulation on the Supervision and Control of Shipments of Waste 259/1993
EU Regulation on Shipments of Waste (1013/2006)
Hazardous Waste (England and Wales) Regulations 2005, SI 894

EU Regulation on Shipments of Waste (1013/2006)

Landfill (England and Wales) Regulations 2002 SI 1559

Landfill (England and Wales) (Amendment) Regulations 2004 SI 1375

Landfill (England and Wales) (Amendment) Regulations 2005 SI 1640

Packaging (Essential Requirements) Regulations 2003 SI 1941

Packaging (Essential Requirements) (Amendment) Regulations 2004 SI 1188

Packaging (Essential Requirements) (Amendment) Regulations 2006 SI 1492

Producer Responsibility Obligations (Packaging Waste) Regulations 2007 SI 871

Transfrontier Shipment of Radioactive Waste Regulations 1993 SI 3031

Transfrontier Shipment of Waste Regulations 2007 SI 1711

Transfrontier Shipment of Waste (Amendment) Regulations 2008 SI 9

Waste Electrical and Electronic Equipment Regulations 2006 SI 3289

Waste Electrical and Electronic Equipment (Amendment) Regulations 2007 SI 3454

Waste Electrical and Electronic Equipment (Waste Management Licensing) (England and Wales) Regulations 2006 SI 3315

Waste Electrical and Electronic Equipment (Waste Management Licensing) (England and Wales) (Amendment) Regulations 2007 SI 1085

Waste Incineration (England and Wales) Regulations 2002 SI 2980

Waste Management (England and Wales) Regulations 2006 SI 937

Waste Management Licences (Consultation and Compensation) Regulations 1999 SI 481

Waste Management Licensing Regulations 1994 SI 1056

Waste Management Licensing (Amendment) Regulations 1995 SI 288

Waste Management Licensing (Amendment No. 2) Regulations 1995 SI 1950

Waste Management Licensing (Amendment) Regulations 1996 SI 1279

Waste Management Licensing (Amendment) Regulations 1997 SI 2203

Waste Management Licensing (Amendment) Regulations 1998 SI 606

Waste Management Licensing (Amendment) (England) Regulations 2002 SI 674

Waste Management Licensing (Amendment) (England) Regulations 2003 SI 595

Waste Management Licensing (England and Wales) (Amendment and Related Provisions) (No. 3) Regulations 2005 SI 1728

Waste Management Regulations 1996 SI 634

Waste Management (Miscellaneous Provisions) (England and Wales) Regulations 2007 SI 1156

References

1. http://www.environment-agency.gov.uk/commondata/103196/ 1753207?referrer=/subjects/waste/1031954/315439/1720716/.
2. http://www.landfill-gas.com/html/landfill_gas_explosions.html.
3. IPCC Third Assessment Report, IPCC Fourth Assessment Report.

4. http://www.landfill-gas.com/html/energy_from_waste.html.
5. *The Digital Dump: Exporting Re-Use and Abuse to Africa* – BAN Report 2005.
6. Council Directive 1999/31/EC of 26 April 1999 on the landfill of waste.
7. P. T. Williams, *Waste Treatment and Disposal*, John Wiley & Sons, 1998.
8. http://www.ciwem.org/resources/waste_incineration.asp.
9. Council Directive 89/429/EEC of 21 June 1989 on the reduction of air pollution from existing municipal waste-incineration plants.
10. Council Directive 89/369/EEC on the prevention of air pollution from new municipal waste incineration plants.
11. http://www.ciwem.org/resources/waste_incineration.asp.
12. Ethos to produce energy from household waste – http://www.ethosrecycling.co.uk/news.html, accessed 22nd January 2008.
13. Graveson Energy Management Ltd (GEM), http://www.gem-ltd.co.uk/?cat=5&id=28.
14. Stockholm Convention on Persistent Organic Pollutants (POPs), May 2004, http://www.pops.int/documents/convtext/convtext_en.pdf.
15. D. Dockery *et al.*, *Environ. Health Perspect.*, 1996, **104**(5), 500–505.
16. Health Effects Institute (HEI), Particulate Air Pollution and Daily Mortality: Replication and Validation of Selected Studies, HEI, Cambridge, MA, 1995.

Recycling and Recovery

DARREN KELL

1 Introduction

Recycling and recovery of e-waste is concerned with the separation of individual materials (copper, steel, polystyrene, *etc.*) into grades that are saleable on the open market or to other secondary processors. Large-scale recycling started in the 1950s as large telephone exchanges and other communications equipment began to be replaced by new technology. This early recycling was driven by economics as people realised the inherent value of the copper and other metals contained in these facilities. Only the high-value metals were of interest to early recyclers and this has continued until very recently. This has led to well-developed technology to separate metals that is capable of extremely high recovery rates, whereas the technology to separate and recover plastics is relatively underdeveloped. The introduction of legislation has meant that significant effort is now going into the recovery of lower-value materials, particularly plastics, in order to meet the legislative targets.

Japan was one of the first countries to legislate for e-waste recycling through the Home Appliances Recycling Law (HARL), enacted in 2001. HARL covers only the four largest groups of appliances (TVs/monitors, refrigeration equipment, air-conditioning equipment and washing machines); recycling rates in Japan are between 64 and 84% depending on the category of waste.[1]

The EU introduced the Waste Electrical and Electronic Equipment (WEEE) Directive in 2002; member states then had to implement the directive in national law. This was completed in 2007, the UK being one of the last countries to implement the Directive. Recycling rates in the EU vary from 25 to 40% for medium and large appliances.[2]

The United States has lagged behind in enacting legislation and there is still no national regulation, although some states have implemented their own state

Issues in Environmental Science and Technology, 27
Electronic Waste Management
Edited by R.E. Hester and R.M. Harrison
© Royal Society of Chemistry 2009
Published by the Royal Society of Chemistry, www.rsc.org

regulations. Consequently recycling rates in the US are relatively low with only ~17% of the total waste arisings being recycled in 2005.[3]

The recycling and recovery process is broadly similar throughout the world and generally includes the following steps:

- Separation and sorting (different types of equipment are sorted)
- De-pollution (toxic or environmentally harmful substances are removed)
- Size reduction (equipment is shredded to liberate the different materials)
- Materials separation (*e.g.* ferrous metals removed by magnets)

Many of the processes, particularly size reduction and separation, are repeated iteratively to further upgrade the materials.

2   Separation and Sorting

Some types of equipment are generally collected separately because of the different processes required. Refrigeration equipment containing ozone-depleting substances (CFCs and HCFCs) are collected and stored separately and use the established infrastructure set up after the EU Regulation 2037/00 on ozone-depleting substances.

Televisions and monitors containing cathode ray tubes (CRTs) are also collected separately due to the different process required.

Some small high-value items are collected separately for logistical and economic reasons, *e.g.* postal collection schemes for mobile phones exist because of the small size and high metal values contained in them. Segregated mobile phones are a valuable commodity in the WEEE recycling market.

Non-segregated WEEE must be sorted at the treatment facility to remove equipment which may contain problem materials. Annex II of the WEEE directive lists a number of materials and components that must be removed from collected WEEE (see Table 1). There are two general approaches: manual disassembly or automated 'opening' followed by manual picking of the hazardous components.

The choice of approach to de-pollution is influenced by capacity, larger facilities tending to use less manual disassembly, and also by the preceding collection and transportation steps. Lack of segregation and handling by front loader and bulk containers can preclude any significant manual disassembly because of the level of breakage during transport. The level of segregation of different materials, and how completely hazardous materials are removed, is heavily dependent on the operators. Training of operators is key to the effectiveness of de-pollution and segregation (Figures 1–4).

3   Treatment

Different processes exist for different types of equipment; CRTs and fridges/freezers are dealt with in specialised dedicated facilities to cope with the

Table 1   Components to be removed, Annex II of the WEEE directive.

As a minimum the following substances, preparations and components have to be removed from any separately collected WEEE:

- polychlorinated biphenyls (PCB) containing capacitors in accordance with Council Directive 96/59/EC of 16 September 1996 on the disposal of polychlorinated biphenyls and polychlorinated terphenyls (PCB/PCT)(1),
- mercury-containing components, such as switches or backlighting lamps,
- batteries,
- printed circuit boards of mobile phones generally, and of other devices if the surface of the printed circuit board is greater than 10 square centimetres,
- toner cartridges, liquid and pasty, as well as colour toner,
- plastic containing brominated flame retardants,
- asbestos waste and components which contain asbestos,
- cathode ray tubes,
- chlorofluorocarbons (CFC), hydrochlorofluorocarbons (HCFC) or hydro-fluorocarbons (HFC), hydrocarbons (HC),
- gas discharge lamps,
- liquid crystal displays (together with their casing where appropriate) of a surface greater than 100 square centimetres and all those backlit with gas discharge lamps,
- external electric cables,
- components containing refractory ceramic fibres as described in Commission Directive 97/69/EC of 5 December 1997 adapting to technical progress Council Directive 67/548/EEC relating to the classification, packaging and labelling of dangerous substances(2),
- components containing radioactive substances with the exception of components that are below the exemption thresholds set in Article 3 of Annex I to Council Directive 96/29/Euratom of 13 May 1996 laying down basic safety standards for the protection of the health of workers and the general public against the dangers arising from ionising radiation(3),
- electrolyte capacitors containing substances of concern (height >25 mm, diameter > 25 mm or proportionately similar volume).

These substances, preparations and components shall be disposed of or recovered in compliance with Article 4 of Council Directive 75/442/EEC.

hazardous or environmentally damaging substances present. Most other WEEE is processed in a mixed WEEE facility.

3.1   Mixed WEEE

Most mixed WEEE is processed in a facility based on mechanical separation; the overall process is outlined in Figure 5.

Many of the processes, particularly size reduction and separation, are repeated iteratively to further upgrade the materials. Size classification and dust removal are also important in the separation process with many facilities operating parallel processes on different size fractions. Manual sorting is also a common operation used to remove large identifiable items such as circuit boards, stainless steel and cables.

Figure 1   Segregated WEEE collection.

Figure 2   Manual disassembly.

The number of iterations, the quality of manual sorting and the operation and layout of the plant (*e.g.* in order to ensure an even layer of material into the separation processes) all have a large impact on the efficiency of the process and quality of the output materials.

Figure 3   Bulk mixed WEEE.

Figure 4   Manual picking line.

3.2   *Refrigeration Equipment*

Refrigeration equipment is processed in dedicated facilities, as required by EU Regulation 2037/00. A number of processes exist but are broadly similar; a typical process is outlined in Figure 6. After some manual disassembly and de-pollution (typically removal of compressor oil and refrigerant) units are

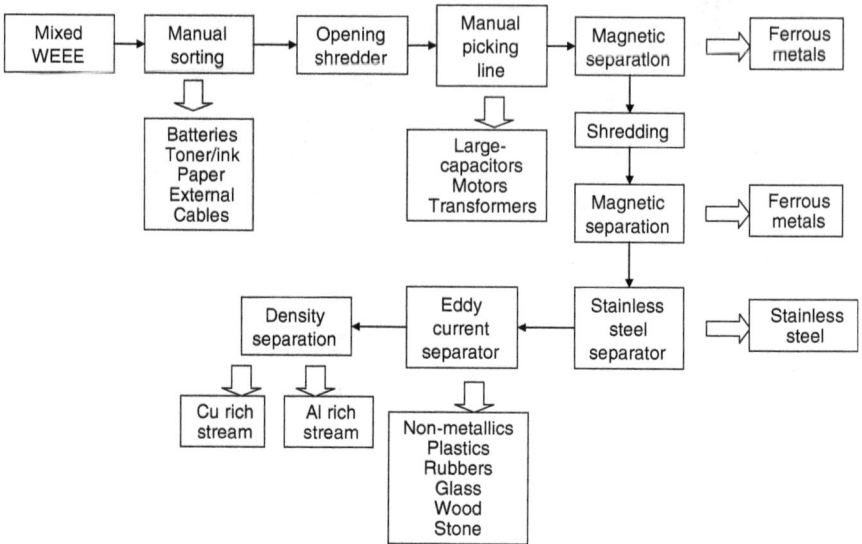

Figure 5   Simplified process flow for a mixed WEEE metal recovery plant.

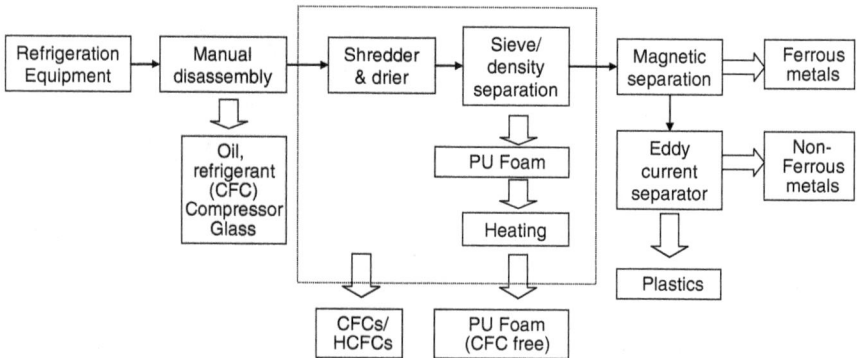

Figure 6   Simplified process flow for refrigeration equipment.

loaded into a sealed system where the blowing agent contained in the insulation and any residual refrigerant can be contained and separated. Units are shredded and the foam insulation heated to drive off the blowing agent (CFC/HCFCs). The CFC/HCFCs are drawn out of the system and condensed either for disposal or for further upgrading. The subsequent metal recovery processes use the same technology as for mixed WEEE.

3.3   Cathode Ray Tubes

Equipment containing CRTs is processed in dedicated facilities because of the need to contain and remove the phosphor coating inside the tube and to deal

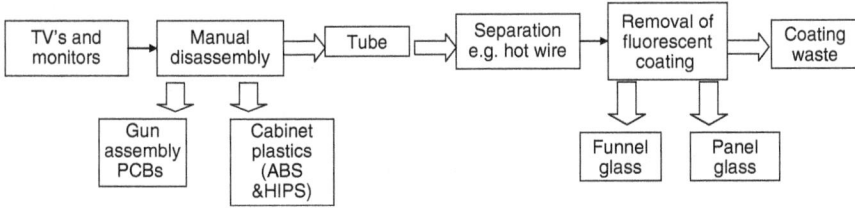

Figure 7 Simplified flow sheet for CRT treatment.

with other hazardous materials in the glass and electrodes. After the tube is manually removed from the equipment there are a number of processes for treating the tube itself. The majority of facilities separate the panel glass from the funnel glass then manually remove the inside coating. A number of methods exist to separate the panel and funnel, *e.g.* using a hot wire or diamond saw. Some facilities crush the whole CRT and wash the coating out before separating the two types of glass in an automated system, *e.g.* by density. A typical process is outlined in Figure 7.

3.4 Individual Processes

3.4.1 Crushing/Diminution

One of the most common pieces of equipment used for initial crushing and shredding is a hammer mill. Hammer mills accomplish size reduction by impacting a slow moving target with a rapidly moving hammer. The target has little or no momentum (low kinetic energy), whereas the hammer tip is travelling at rates of typically up to $7000 \, \text{m min}^{-1}$ and higher (high kinetic energy). It is the transfer of energy resulting from this collision that fractures the feedstock. Sizing is a function of hammer speed, hammer design and placement, screen design and hole size and air assist. Because impact is the primary force used in a hammer mill to generate particulate reduction from bulk, anything that increases the chance of a collision, increases the magnitude of the collision or enhances material take-off is advantageous to particle size reduction. The magnitude of the collisions is normally increased by increasing hammer speed; this produces particles of smaller mean geometric size; see Figure 8.

Material disintegration may also be effected by the use of metal crushers which have low specific energy consumption and offer high operational immunity to the presence of solid pieces and may be also used as a pre-stage prior to shredding; see Figure 9.

3.4.2 Size Classification

Screeners are sifting units that are rotated as powder is fed into their interior. The finer particles fall through the sieve opening and oversized particles are ejected off the end. Rotary sifters or drum screeners are often used for

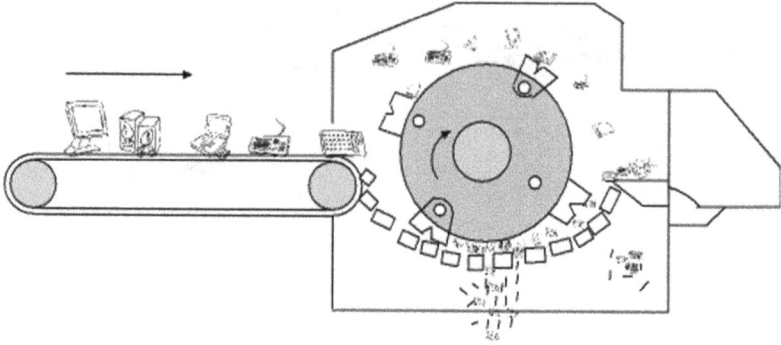

Figure 8   Hammer mill.

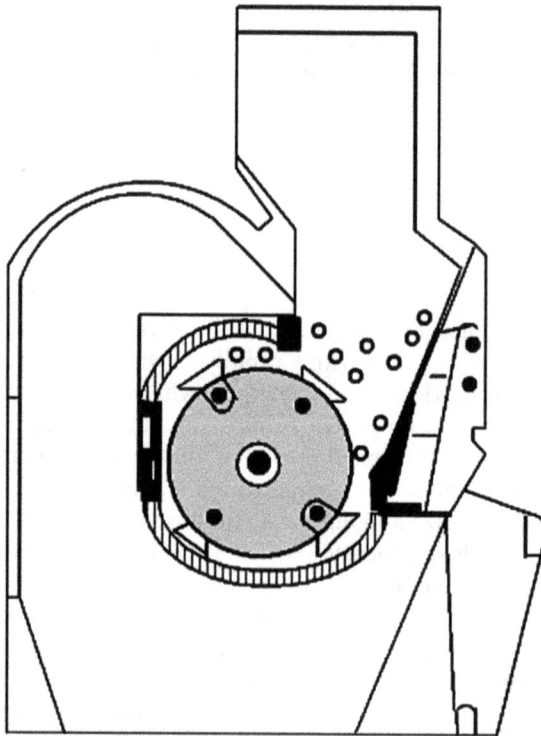

Figure 9   Crusher.

de-agglomerating or de-lumping type operations. Screeners are available in three main types: drum sifter, rectangular deck and round deck.

Air classifiers, cones or cyclones use the spiral air flow action or acceleration within a chamber to separate or classify solid particles. Powders suspended in air or gas enter the cyclone and the heavier particles spiral out and down where they are collected. The air and finer particles flow up to the top where they may be passed to another cyclone with finer classification capability. A cyclone is essentially a settling chamber where the effects of gravity (acceleration) have been replaced with centrifugal acceleration.

Concentrating tables or density separators screen bulk materials or minerals based on the density (specific gravity), size and shape of the particles. This group includes jigging equipment, hindered-bed settling devices, shaking tables, spiral concentrators, concentrating or wet tables, hydraulic concentrating tables, constriction plate separators or specialized settling vessels. Most concentrating or density separation equipment is hydraulic or water-based, although pneumatic or air-based systems are also available.

Trommels are large rotary drums with a grate-like surface with large openings. Trommels are used to separate very coarse materials from bulk materials such as coarse plastics from finer aluminum recycled material, coarse inorganic materials from organic wastes or large ore chunks from finer minerals.

Water classifiers, such as elutriators and classifying hydrocyclones, use settling or flow in water or another liquid to separate or classify powdered materials on the basis of particle size or shape.

3.4.3 Magnetic Separation

Magnetic separators such as low-intensity drum types are widely used for the recovery of ferromagnetic materials from non-ferrous metals and other non-magnetic materials. There have been many advances in the design and operation of high-intensity magnetic separators due mainly to the introduction of rare-earth alloy permanent magnets with the capability of providing high field strengths and gradients. For WEEE, magnetic separation systems utilise ferrite, rare-earth or electromagnets, with high-intensity electromagnet systems being used extensively and which are particularly suited to materials fed to the underside of drum magnets; see Figure 10.

3.4.4 Density Separation

Several different methods may be deployed to separate heavier fractions from lighter ones, the basis being the difference in density to enable such. Gravity concentration separates materials of different specific gravity by their relative movement in response to the action of gravity and one or more other forces, such as the resistance to motion offered by water or air. The motion of a particle in a fluid is dependent not only upon the particle's density, but also on its size and shape; large particles are affected more than smaller ones. In practice, close size control of feeds to gravity separation equipment is required in order to minimise size effects and render the relative motion of the particle gravity dependent.

The use of air to separate materials of different density has long been established and air tables are used extensively within the food industry for grain separation and within the metals industry for applications such as refining of crushed slag in foundry output. In recent years both air and water-based

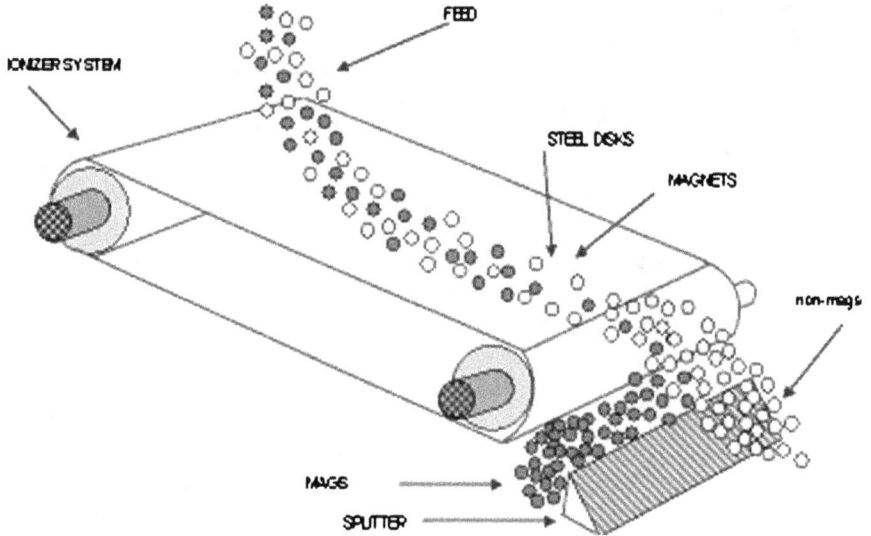

Figure 10   Magnetic separator.

gravity tables have been adapted for the sorting of electronic scrap and form an integral part of a number of electronic recycling plant operations.

Essentially, an air gravity table is similar to a mechanised gold pan that operates continuously and with a high degree of efficiency. The table is comprised of a deck, in somewhat of a rectangular shape, covered with riffles (raised bars running perpendicular to the feed side of the table), mounted in a near flat position, on a supporting frame that allows the table to slide along the long axis of the table. Instead of water as the medium, air is used and is continuously injected through the porous bed of the table.

The mechanism is attached to the table, and it moves the table along the long axis a distance adjustable between 1/2 and 1 inch and then back to the starting position between 200 and 300 times per minute. This reciprocal movement is faster on the reverse stroke than it is on the forward stroke. This shaking movement helps transport the concentrates or heavy material to the concentrate end of the table. A very important operating variable of a shaking table is the tilt adjustment. Normally, the feed side is lower, and the concentrate end is higher on an air table, which creates an upward slope where the heavy material will ascend, while the light density material will not, and, consequently, will flow over the riffles. The tailing (low density) side is near level to lower than the feed side. Another important variable in air table operation is the volume of air, and this is typically adjusted by a series of valves, or plate type regulators, allowing more or less air to flow to the deck. It is important to have a uniform flow of air across the deck.

Feed is introduced to the feed box in a narrow size range. For air tables to function effectively, the feed needs to be in a narrow size range, usually within a ratio of 2.8:1, from the smallest particle to the largest particle. The maximum particle size is about 3.2 mm and the typical fine size that can be separated on an air table is normally 0.25 mm. If 0.25-mm-sized material is to be separated, the feed size range would be from 0.69 mm to 0.25 mm. In all gravity separations, a difference in specific gravity of the materials needs to be significant, at 1 or greater (*e.g.* a 2.2 SG material will usually separate from a 3.2 SG material). Air tables can also separate, somewhat, on particle shapes, as differing particle shapes react differently in the rising air columns. Generally, these will show up in the middling (discharge between tailings and concentrate).

The riffles are always taller on the feed side of the table, and decrease in height as they progress towards the tailings side of the table. This allows for the quick separation of the larger high-density material, and allows more residence time for the more difficult, finer, high-density particles to separate from the finer, low-density material.

Air tables work in a similar way to wet gravity tables, in that the material is fed perpendicular to the riffles, and the high-density material remains behind the riffles, and the fluidising air columns rise through the bed of material, relative to Stokes' law and Hindered Settling. This causes settling at differential rates, where the low-density material flows above and over the riffles, while the high-density material stays close to the deck surface, and follows the riffles to the concentrate discharge.

3.4.5 Eddy Current Separation

When particles of non-ferrous metals pass over a rotating magnet at high speeds, eddy currents are created in the non-ferrous metal, which generates a magnetic field around them. When the polarity of that magnetic field is the same as the rotating magnet, the non-ferrous metal is repelled from the magnet and it is this action that enables separation of such metals from feedstock material. Practical deployment of an eddy current separator is generally by action on granulated material; the material is transported on a conveyor belt which moves over a pulley which houses a high-intensity permanent magnet rotating at high speed, thus generating an alternating magnetic or primary magnetic field. An electric current (eddy current) is generated in conductors which are exposed to this field, with a resultant repulsive force existing between the drum magnet arrangement and the conductor which in turn moves the conductor away from the drum. The generation of such a lifting force affects the trajectory of particles leaving the belt; this affords the practical means of separation of non-ferrous particulates. Typical particulate sizes processed tend to be in the 3 to 150 mm size range.

The introduction of eddy current separators, whose operability is based on the use of rare-earth permanent magnets, has been one of the most significant developments in the recycling industry in recent years. The separators were initially developed to recover non-ferrous metals from shredded automobile scrap or for the treatment of municipal solid waste. They are particularly suited to the handling of relatively coarse-sized feeds. High-frequency eddy current separators, where the magnetic field changes very rapidly, are needed for separation of smaller particles.

3.4.6   Electrostatic Separation

The rotor type electrostatic separator, using corona charging, may be utilised to separate raw materials into conductive and non-conductive fractions. The extreme difference in the electrical conductivity or specific electrical resistance between metals and non-metals affords an excellent pre-condition for the successful implementation of a corona electrostatic separation in recycling of waste. Electrostatic separation has been mainly used for the recovery of copper or aluminium from chopped electric wires and cables and, more specifically, for the recovery of copper and precious metals from printed circuit board scrap; see Figure 11.

4   Outputs and Markets

Downstream markets for the outputs of WEEE recyclers are complex and often consist of further layers of 'recyclers' rather than end markets in themselves. The fact that WEEE processors vary widely in their outputs adds to this complexity, *e.g.* some facilities upgrade the plastics fraction to give a cleaner stream containing fewer polymer types, but many do not. There are some standards for trading in waste electronics that have undergone various degrees of processing, *e.g.* those set by the Institute of Scrap Recycling Industries.[4] Many processors have individual specifications with their customers.

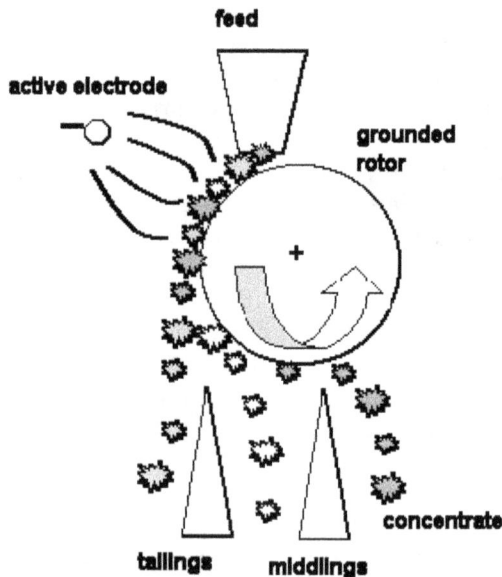

Figure 11   Electrostatic separator.

4.1 Metals

As noted previously, metals constitute the most valuable and easiest-to-recycle materials. A recent study concluded that 'There appear to be no major difficulties concerning the recovery and recycling of metals from WEEE . . . There are ample capacities and markets available.'[2] Metals also constitute the largest weight of materials in WEEE, around 47% overall for small mixed WEEE.[5] Current recycling processes are capable of recovering <95% of the in-feed metals; see Table 2.

4.2 Glass

The major source of glass in waste electronics is from CRTs, although this is likely to shift towards glass from flat-panel displays as the number of CRTs declines. CRTs are composed of two main glass types: funnel glass (the back of the tube), which contains high levels of lead oxide, and panel glass (the screen), which contains barium and strontium oxides. In the UK most of the CRT glass is not recycled.[6] The preferred route for recycling is a closed-loop system, using the glass in the manufacture of new CRTs. This is difficult for a number of reasons: (i) low levels of contamination are required for the production of new tubes that are difficult to reach with post-consumer recycled product; (ii) there is a declining market for CRTs as they are replaced by flat panel displays; (iii) manufacture of CRTs is now mainly carried out in non-OECD countries, making the transport of 'waste' glass difficult because of the transfrontier shipment of waste regulations.

A number of other outlets for recycled glass have been identified, including use in bricks and other building products, as aggregate and as a flux in smelting operations.[6] These have so far not achieved widespread commercial use.

4.3 Plastics

On average plastics constitute approximately 20% of collected WEEE.[7] A number of distinct materials streams are produced from WEEE recycling. Some

Table 2 Metals recycling.

Material	Source	Market
Ferrous metals	Magnetic separation	Reducing grade scrap to steel industry
Copper (precious metals are also contained in the PCBs and copper fraction)	Hand sorting of PCBs, copper fraction from density separator, cables	Copper smelters
Stainless steel	Hand sorting of washing machine drums, for example; rare-earth magnet separation	Stainless steel production
Aluminium	Non-ferrous density separation	Aluminium production

from segregated appliance types are fairly homogeneous and can be recycled. The plastics stream from mixed WEEE processing contains a large number of different polymers that are difficult to separate – according to a recent study 'Data from literature seems to confirm that at present plastic output streams from WEEE recycling operations are mostly not recovered, but are landfilled together with other residue streams'.[2]

Television and monitor casings are usually collected separately as part of the dismantling operation necessary to treat the CRT. These casings are predominantly a mixture of High-Impact Polystyrene (HIPS) and Acrylonitrile Butadiene Styrene (ABS), which can be used as a mixture in injection moulding or further separated. Currently the high incidence of brominated flame retardants (BFRs) in this material makes recycling difficult in the EU because of regulation.[8,9] In Japan the material is used in low-grade applications where fire retardance is a benefit.[1]

Plastics from refrigeration equipment are also collected separately from dedicated processes; again this is a mixture of HIPS and ABS. Being free from BFRs, this material can be recycled.

The composition of plastics from mixed WEEE processing is complex, containing at least five different polymers in large amounts and many more used in smaller quantities for specific applications.

The major components are:

- Acrylonitrile Butadiene Styrene (ABS)
- Polycarbonate (PC)
- PC/ABS blends
- High-Impact Polystyrene (HIPS)
- Polyphenylene Oxide blends (PPO)

Currently most of this mixed plastic is not recycled; depending on the facilities in the country, it is either landfilled or incinerated. Some mixed WEEE plastics are exported, particularly to China, where the material is sorted by hand into different polymer types; this practice is unlikely to be sustainable in the long term for both social and economic reasons.[10]

5 Emerging Technologies

5.1 Separation

Efficient separation is a prime requirement for effective WEEE recycling and can reduce reliance on dismantling. Although new techniques are being developed, much of the novelty in WEEE separation comes from adapting existing techniques and from novel combinations. Initially this is usually for specific input streams, but more sophisticated routes are being developed to handle a wider variety of input items. Current technologies can separate plastics from other materials, but segregation of the different types is a key aim, in which sensors have a vital role. Mechanical devices for plastics classification include a novel air force sorter, and a 'weak flow watercourse separator'. Adherent technologies in the US reported a complete 'conversion process' that gasifies the plastics in heated screw feeds, and then reforms them into secondary materials, while metals are recovered by leaching, electrowinning and electrorefining.[11]

Some researchers are combining separation techniques (*e.g.* screw threads within floatation tanks, jiggers or conveyor belts with water added) in some cases with superimposed magnetic fields.[12] In other areas, existing techniques are being enhanced, *e.g.* eddy current methods with rotating electrodes or with pulsed excitation, and corona separators with novel electrode designs.

Electrostatic separators are good for extracting plastics, but are limited to relatively small particle size. Sand-based fluidised beds for gravity separation are also under investigation. Value may be extracted even from residual powder materials in processes under research in the automotive industry. Following pyrolysis and magnetic separation, sufficient carbon and iron dust is retrieved to render the process viable.[13]

A major advance in separation is likely to come through sensing methods. Opto-electronic sorting[14] is now being incorporated into research systems. The construction sector is developing electromagnetic field methods for both sensing and sorting. At low field levels, enhanced with pulsed excitation, signature combinations of conductivity and permeability of each material allow them to be discriminated. Higher fields eject the metal pieces from a conveyor belt while signals are fed to a processor that activates appropriately positioned air nozzles to effect separation in the falling zone.[15]

Another system under research in the construction sector uses a similar intelligent ejection unit, but linked to a camera, and sorting on colour, shape and position.[16] X-ray systems are also under investigation, and could equally be linked to such an automatic ejection system.

5.2 Thermal Treatments

Thermal treatments have the advantages of greatly reducing bulk and avoiding liquid effluent for the primary recycler, although ultimately further refining is necessary to extract pure metals. They are not yet widely implemented in recycling of whole appliances, but are subject to much research, often adapting methods from the mining industry. Pyrometallurgical routes are potentially suitable for PCBs, which contain 29%wt of valuable metals, even though comprising only 3%wt of WEEE. Typically, after some initial sorting, the organic content is reduced to ash, which can be used as feedstock in the pyrometallurgical processing. The final products tend to be loosely refined metal ingots, such as ferrous, aluminium, mixed Pb/Sn and, most importantly, a copper-rich precious metals mix. These are suitable for further refining. There are established operators (*e.g.* Engelhard and Boliden) and much continuing research into these methods, *e.g.* partial vacuum methods are being investigated. Other thermal routes include encapsulation, using either glass or binder, to produce low-grade block products for use in construction. Harmful contents are safely sealed in, but the opportunity to recycle valuable resources is lost. Research topics include use of vacuum, thermal plasmas and lasers to enhance the thermal treatment. Other emerging variants include processes for copper-rich and iron-rich items. The latter is adapted from the car industry, but claimed to be suitable for WEEE.

5.3 Hydrometallurgical Extraction

Hydrometallurgy is well established for extracting and applying precious metals. The need to reduce reliance on the hazardous cyanide containing solutions that are routinely used drives current research.

Processes based on strong acids and hydrogen peroxide have been developed for WEEE,[17] with fluoroboric acid proving useful for extraction from mixed streams, including products from pyrolytic processes. The search for still less hazardous reagents continues, however. Thiourea and thiosulfate are candidates,[18] as well as solvents such as polyhydric alcohol, ketones, polyether or cyclic lactone.[19] Stability, process control, reagent recyclability and economics are the outstanding issues for such methods. Research in the mining industry has generated a methodology based on oxygen, ozone and a complexing agent, which facilitates passivation of the waste feed-material and selective recovery of metals, without by-products.[20] Catalytic processes are being developed for tin and lead.

Electrochemical methods are under investigation. A novel combined approach in which a non-selective leach process permits selective recovery in a subsequent electrochemical reactor is under research at Imperial College.[21] Further novelty in combined approaches is found in a project by the University of Birmingham, Alchema and C-Tech Innovation, where selective electro-chemical recovery is augmented by microwave-enhanced biodigestion for the precious metals.[22]

5.4 Sensing Technologies

Sensing methods can greatly improve the effectiveness of WEEE recycling. They are crucial to implementation of automated disassembly and can facilitate great improvements in separation. Opto-electronic sorters, which use conventional imaging devices to discriminate on shape and colour, have been developed for various industries. Augmentation by electromagnetic sensing permits identification of metals, as well as of rubbers and plastics,[23] allowing selective ejection of the identified items in automated separation processes. Laser Induced Breakdown Spectroscopy (LIBS) is a laboratory technique that is being adapted for on-line operation in separation processes, and to which enhancements such as pulsing are being applied. It is useful for heavy metals, and is of particular interest because it can detect brominated flame retardants. High accuracy and operational speed have been demonstrated.[24] Precise identification of plastics is key to economic viability of WEEE recycling. Current recycling methods generate low-grade mixed plastics, suitable only for de-rated applications. Laboratory analytical techniques that are under research for adapting to this application include laser-induced fluorescence[25] and X-rays.[26] Adaptations of electromagnetic sensing for conducting materials are also being developed.

A laser-based system,[27] originally developed for the food industry, which detects a combination of shape, colour and reflectance, has recently been trialled for separation of various WEEE waste streams including mixed plastics, with encouraging results.

5.5 Plastics to Liquid Fuel

A number of processes have been proposed to use the mixed plastic fraction as a source of organic chemicals that can be converted to a liquid fuel. The most developed of these is the Catalytic Depolymerisation Process (CDP)[28] invented by Dr Christian Koch. Suitable feed materials for the CDP process are claimed to be almost any hydrocarbon-based material from mixed plastic to tar lakes. Material must be dry, less than 10 mm (if solid) and free of metals and other inert materials (glass, stone, *etc.*). PVC is transformed into non-toxic salts like calcium chloride. Chalk is added to the reactor which also regenerates the catalyst. The inventor says the process does not produce dioxins or furans.

The pumps are one of the key parts of the process. Heating is carried out in the pump, rather than in the large vessel. This ensures that no part of the fluid is heated to the point at which dioxins are produced ($> 400\,^\circ$C). The pump uses friction to heat the reactants while bringing together the reactants and catalyst into intimate contact.

Catalyst consumption is around 3% of the diesel produced by weight. The process is claimed to be around 90% efficient, *i.e.* 90% of the calorific value of the input material is available from the diesel produced. The energy required to run the plant is equivalent to 10% of the diesel produced therefore the process should be around 80% efficient overall.

The first operational plant is located in Mexico and can produce *c.* 500 l hour^{-1} of diesel. A second, larger, plant capable of 1500 l hour^{-1} is planned for Canada. The design has been adapted to be modular so it can be easily transported as standard containers and easily installed on site. This also makes it possible to transport the plant to the waste rather than the other way round.

5.6 Plastics Containing Brominated Flame Retardents

A number of attempts have been made to remove BFRs from plastics to allow recycling of the plastic.[29] One such process is the Creasolv™ process, developed in co-operation by the Fraunhofer-Institute for Process Technology and Packaging IVV, Freising and CreaCycle GmbH, Grevenbroich, Germany.

It is based on selective extraction of a targeted polymer from plastic waste, followed by a cleaning step. Impurities, undesired additives (*e.g.* flame retardants) and toxic degradation products can be separated effectively to obtain a high-purity polymer. The tar residue from this process is rich in bromine. If the bromine level is >10% then it can be used as a feedstock into the bromine industry, this is a particularly good example of how refining and purification of a waste stream adds to the value, and how almost any substance can have value, even toxic substances.

6   Acknowledgements

Some of the material presented here was originally included in a report for the DEFRA-funded WEEE-Tech project. Figures 8–11 courtesy of Rod Kellner, one of the Authors of the reports.

References

1. Global Watch Mission Report, *Waste Electrical and Electronic Equipment (WEEE): Innovating Novel Recovery and Recycling Technologies in Japan*, September 2005, URN 06/510.
2. 2008 Review of Directive 2002/96 on Waste Electrical and Electronic Equipment (WEEE), Study No. 07010401/2006/442493/ETU/G4, United Nations University.
3. US Environmental Protection Agency, Management of Electronic Waste in the United States, EPA530-D-07-002, November 2007.
4. Institute of Scrap Recycling Industries Inc., Scrap Specifications Circular 2007, Washington.
5. DEFRA, *Trial to establish Waste Electrical and Electronic Equipment (WEEE) Protocols*, 2007.
6. ICER (Industry Council for Electronic Equipment Recycling), *Materials Recovery from Waste Cathode Ray Tubes (CRTs)*, WRAP, Banbury, 2004, ISBN: 1-84405-077-7.
7. S. Wilkinson, N. Duffy, M. Crowe and K. Nolan, *Waste from Electrical and Electronic Equipment*, Environmental Protection Agency, Ireland, May 2001.
8. Directive 2002/96/EC of the European Parliament and of the Council of 27 January 2003 on waste electrical and electronic equipment (WEEE).
9. Directive 2002/95/EC of the European Parliament and of the Council of 27 January 2003 on the restriction of the use of certain hazardous substances in electrical and electronic equipment.
10. C. Hicks, R. Dietmar and M. Eugster, *Environ. Impact Assess. Rev.*, 2005, **25**(5), 459–471.
11. L. D. Busselle, T. A. Moore, J. M. Shoemaker, R. E. Allred, Proc. *IEEE International Symposium on Electronics and the Environment*, Danvers, MA, USA, 11–13 May 1999, vol. xii, pp.192–197, ISBN: 0-7803-5495-8.

12. Matsushita Denki Sangyo KK, Sorting device for recycling of domestic electrical appliances, separates non-ferrous material, resin and lightweight materials from crushed raw material, based on their specific gravities, JP2002086013 A UPAB: 20020916.

13. Toyota Jidosha KK, Processing method, for recovering iron and carbon powder from scrap car and waste electrical appliance, involves classifying magnetic and non-magnetic components in waste material using magnetic force classification, JP 2003320360 A 20031111 (200382).

14. Metal 'X' Technology for Automated Metal Sorting, http://www.commodas. de/holding/recycling/metal.html.

15. M. B. Mesina, T. P. R. de Jong, W. L. Dalmijn and M. A. Reuter, *Materials*, March 14–18, 2004, Charlotte, North Carolina, USA.

16. T. P. R. De Jong and L. Fabrizi, *Recycling International*, December 2004.

17. S. Hermann and U. Landau, Finest refinement of gold containing metallic impurities, comprises dissolving in hydrochloric acid and hydrogen peroxide, precipitating and washing, EP 1606424, WO 2004081245.

18. J. Kulandaisamy, J. Prabhaker Rethinaraj, P. Adaikkalam, G. N. Srinivasan and M. Raghavan, *JOM*, 2003, **55**(8), 35–41.

19. M. Ishikawa, Y. Kawase, F. Mizutani and T. Sakakibara, Solution for recovering noble metals such as gold, silver and platinum from electronic device, contains iodine and/or iodide ion and organic solvent selected from polyhydric alcohol, polyether or cyclic lactone, JP 2005154892.

20. J. Casado Gimenez, M. Cruells Cadevall, E. Juaon Morera, A. Roca Vallmajor and J. Vinyals Olia, Leaching metals from materials e.g. scrap and ores, involves contacting material comprising metals, with solution containing oxygen, ozone and complexing reagent, EP 1281779 A2 20030205 200323.

21. N. P. Brandon, G. H. Kelsall, T. Muller, R. Olijve, M. Schmidt and Q. Yin, Metal Recovery from Electronic Scrap by Leaching and Electrowinning, 200th Electochemical Soc. Meeting, San Francisco, USA, 2–7 Sept. 2001; *Proc. Electrochem. Soc.*, Energy and Electrochemical Processes for a Cleaner Environment, *Electrochem. Soc.*, NJ, USA, 2001.

22. J. A. Henderson, M. C. Potter, V. S. Baxter-Plant, N. J. Creamer, L. E. Macaskie, W. A. P. Premaratne, N. A. Rowson and A. R. Dale, Novel Bio-Electrochemical Technology for the Recovery of Precious Metals from Electronic Scrap, *Proceedings of the I. Mech. E. conference on 'Profit from Waste VII'*, 27–28 October 2004, One Bird Cage Walk, London, 2004.

23. M. Artinger and H. Frisch, Separator of impurities from plastic or glass stream-uses array of e.g. meta detectors or opto-electronic sensors selectively to actuate nozzles diverting unwanted fragments, EP 353457, 1990.

24. M. Stepputat, R. Noll and R. Miguel, *High-Speed Detection of Additives in Technical Polymers with Laser-Induced Breakdown Spectrometry*, VDI-Berichte, 2002, 1667 (Anwendungen und Trends in der Optischen Analysenmesstechnik), Verlag GmbH, ISSN: 0083-5560.

25. M. Katto and S. Motokoshi, Identification method of kind of plastics for recycling waste plastics from different industries, involves exciting plastic

material using ultraviolet light and using spectroscopic characteristics of its characteristic spectrum distribution, JP 2005164431 A 20050623 (200548).

26. W. L. Dalmijn and T. P. R. de Jong, *Developments in and Applications of Dual Energy X-ray Transmission*, Colloquium Sensorgestützte Sortierung 2004, 6–7 May 2004, Aachen-Stolberg.
27. A detection system for use in a sorting apparatus, a method for determining drift in this detection system and a sorting apparatus comprising such detection system, EP1724029, 2006.
28. C. Koch, Diesel oil from residual materials by catalytic depolymerisation comprising energy input by means of a pump-stirrer system, EP1538191, 2005.
29. K. Freegard, R. Morton and G. Tan, *Develop a Process to Separate Brominated Flame Retardants from WEEE Polymers*, WRAP Oxon., ISBN: 1-84405-315-6.

Integrated Approach to e-Waste Recycling

ROD KELLNER

1 Introduction

The treatment of WEEE may be considered as an integration of a generic waste treatment hierarchy within which economic, legislative and technology drivers determine the structure and methodology of approach. The hierarchy deployed in order of increasing environmental impact may be considered as:

- Reduction
- Re-use
- Recovery
- Recycle
- Disposal

Whilst a reduction in the levels of WEEE in a simplistic sense *via* the extension of product life is clearly the most desirable objective, it should be noted that an impact at this level of the treatment hierarchy may also be realised by such factors as eco-design. In practical terms, re-use (refurbishment/repair) is the least invasive of all options; this would return the item back to its original state, but the end-product may not be of the same quality as the new product. In repair, the product will have components removed and replaced and will be returned, nominally, for the same use as the new product, although the end-specification may be reduced, or indeed enhanced if components of higher specification are installed. Recovery would imply the disassembly of the product with all useful components being recovered and re-used. Recycling may be taken as the product being broken down and the materials recovered to be used in the manufacture of new products. Waste, of course, is the lowest level in the hierarchy and is the level at which no useful components or materials can be

Issues in Environmental Science and Technology, 27
Electronic Waste Management
Edited by R.E. Hester and R.M. Harrison
© Royal Society of Chemistry 2009
Published by the Royal Society of Chemistry, www.rsc.org

obtained and the item must be disposed of by, for example, landfill or incineration.

The following economic factors have a bearing on the optimum end-of-life route:

- **Collection/transport** – Heavily influenced by location, policy, quantity collected *etc.*
- **Sorting** – Mainly manual but also dependent on size of operation and location and policy if additional costs
- **Disassembly** – Currently manual and high costs for one-to-one disassembly – but opportunities for one-to-many disassembly being developed
- **Enforced Processing** – Legislation requires some disassembly/processing and therefore only additional costs need be considered, *e.g.* testing, cleaning, repair
- **Processing for Recovery** – Multiple current options and more being developed to separate and recover valuable or inhibitory components and materials
- **Waste** – Cost of any material not capable of being recycled or of any value
- **Revenue** – Income from re-use or recycling of any products, components or materials

It is important to note the dynamics of any approach to addressing the issues of WEEE and that any derived optimum approach may be subject to change, depending upon varying economic, legislative and technology-driven issues.

In any consideration of a methodology for recovery and recycling is the huge number of variables that need to be taken into consideration. This is particularly true when there are a large number of components even within nominally simple products, and that for similar products there is significant variability. Therefore, it has to be recognised that a definitive solution cannot be provided, as the preferred economic optimum will alter with the variability in the composition of the waste stream, and also the demand for materials being recovered.

For large household white goods, such as fridges and cookers, recycling infrastructure is strong. However, for smaller, more complicated equipment, the development of new infrastructure and technology has become necessary. There are four broad methods employed by industry to recycle:

- **Equipment dismantling** – the manual separation of re-usable and recyclable components
- **Mechanical recycling** – the removal of hazardous components followed by granulating and shredding, in order to remove the recyclable raw materials such as plastic and ferrous metal
- **Incineration and refining** – metal can be recovered after the more combustible material has been incinerated
- **Chemical recycling** – precious metals such as gold and silver can be removed from printed circuit boards and components *via* chemical processes

As in all aspects of WEEE, disassembly is considered an integral element of realising intrinsic value on a scrap printed circuit board (PCB) assembly. As such it is carried out at a number of levels; by the OEM or equipment manufacturers themselves for recovery of components from faulty products or over capacity manufacture for re-use or replacement; by a specialist contractor performing this function for the manufacturer; or by a recycler or disassembler for resale in the secondary component market. Disassembly may also be undertaken as a preparation stage within primary upgrading operations (*e.g.* removal of transformer cores). Practically all such disassembly operations are carried out manually, which in itself places limits on the operation in respect of the costs involved.

Disassembly is considered to be an area of increasing significance in a marketplace of low-cost components and the necessity to address such in an automated low-cost manner. Disassembly may also be considered to have an increasing impact on overall future recycling strategies. It is considered that the limitations of current purely mechanical process recycling routes for PCBs are effectively concerned with precious metal loss from component structures on populated boards (owing to the nature of the metal-to-non-metal interface) and an effective automated disassembly methodology could well expand the potential for mechanical turnkey approaches for all grades of scrap PCBs.

The significance of disassembly is demonstrated with emerging technologies for the recycling of mobile phones. Mobile phones represent a fast-growing WEEE and those assemblies at end-of-life are normally mechanically ground because of the lack of suitable disassembly methods for 'mass disassembling'. The impetus to enhance recyclability is evident in an approach being developed by Nokia whereby the application of magnetic induction heating enables casing cleavage around the screw assemblies with subsequent mechanical impact enabling direct separation into the major components – LCD display, circuit boards and covers.

Whilst there are well-developed recycling technologies being practiced, there are also increasing numbers of emerging techniques which, together with developments to existing techniques, offer the potential to optimise recovery/recycling methodologies on a sectoral basis by, in effect, changing the economic drivers inherent within specific routes. Both existing technologies and emerging technologies are covered in the following section.

2   Recycling and Recovery Technologies

Recycling of WEEE on any reasonable scale is invariably accomplished *via* sequential processing involving:

- Sorting/Disassembly
- Crushing/Diminution – Size Reduction
- Separation
- Recycling

2.1   Sorting/Disassembly

Disassembly Processes

There are a number of methods of disassembly. It is clear that a major cost element within any recycling methodology is that devoted to manual disassembly and sorting, whether this be the manual removal of hazardous components such as batteries and other items prescribed by the WEEE Directive or the manual sorting into classifications such as high- and low-grade material.

Disassembly is a systematic approach that allows the removal of a component, part, group of parts or a sub-assembly from a product (partial disassembly) or the separation of a product into all of its component parts (complete disassembly) for a defined purpose.

Existing practice in the recycling of WEEE places selective disassembly as a vital and integral element of the process in that priority is afforded to the re-use of components, the dismantling of hazardous components and the recovery of valuable materials from printed circuit boards, cables and engineering plastics is simplified by such an approach. Most recycling plants utilise manual dismantling.

The overall process of disassembly may be broadened in definition to include mechanical processes such as physical impacting and primary forms of shredding and fragmenting and, in certain instances, granulation may be interpreted as being within the scope of disassembly.

Physical impaction would comprise methods which break down products to enable the salvaging of re-usable and recyclable parts, components and materials, whereas shredding in its primary form would be the breakdown of the product into pieces *via* fragmentation, ripping or tearing, which may then be sorted into differing material streams which would have dissimilar subsequent processing demands.

Shredding/Fragmenting: a process in which products are fed into a shredder which fragments, grinds, rips or tears the product into pieces, which are then sorted into different materials streams and recyclable or valuable materials extracted.

Granulating: this is the mechanical processing of production scrap, post-consumer plastic packaging, industrial parts or other materials into fine particles.

Process: 'Granulators consist of a feed hopper, cutting chamber, classifying screen and rotating knives that work in concert with stationary-bed knives to reduce the plastic scrap until it is small enough to pass through the classifying screen. The resulting particles, called regrind, can vary in size from 3 mm to 20 mm.'

Mechanical recycling of plastics involves melting, shredding and granulation of waste plastics which must be sorted prior to mechanical recycling into polymer types and/or colour. The plastic would then be melted down directly and moulded into a new shape or melted down after being shredded into flakes and then processed into granules called regranulate. Granulating is ideal for products with a high plastic content such as small household or gardening appliances and power tools. It is important that there are no contaminates, *i.e.* hazardous materials/components, present; therefore these may have to be removed manually prior to granulating. A specialist granulator is used to process WEEE so that different material streams can be easily separated.

Automated Disassembly

Automated disassembly would comprise a fully automated production-line-based disassembly system. Whilst, in principle, automated disassembly would seem to provide a cost-effective means of recovering and recycling components and materials on a large scale, there are advantages and disadvantages to this approach.

Advantages of automated disassembly would be its being ideally suited for high-metal-content waste and its being the cheapest option in respect of minimised labour content when established. Disadvantages, however, would include difficulties in effecting total disassembly of most electronic goods with automation due to problems associated with fasteners and access to parts situated in close proximity, as well as the necessity to work with large quantities of similar product feedstock, the generation of generally cruder fractions than mechanical disassembly and the necessity to still remove hazardous materials as a pre-stage demand.

2.2 Crushing/Diminution

One of the most common pieces of equipment used for initial crushing and shredding is a hammer mill. Hammer mills accomplish size reduction by impacting a slow moving target with a rapidly moving hammer. The target has little or no momentum (low kinetic energy), whereas the hammer tip is travelling at rates of typically $7000 \, \text{m} \, \text{min}^{-1}$ and higher (high kinetic energy). It is the transfer of energy resulting from this collision that fractures the feedstock.

Material disintegration may also be effected by the use of metal crushers which have low specific energy consumption and offer high operational immunity to the presence of solid pieces and may be also used at a pre-stage prior to shredding.

2.3 Separation

General Methodologies

Screeners, classifiers, shakers and separators are used for classification of powders or other bulk materials by particle size as well as separation of particles by density, magnetic properties or electrical characteristics. Round and rectangular screeners, magnetic separators, electrostatic separators, rotary sifters, wet or concentrating tables, rake classifiers, classifying hydrocyclones, floatation systems and trommels are included in the category.

Screeners are sifting units that are rotated as powder is fed into their interior. The finer particles fall through the sieve opening and oversized particles are ejected off the end. Rotary sifters or drum screeners are often used for de-agglomerating or de-lumping type operations. Screeners are available in three main types: drum sifter, rectangular deck and round deck.

Air classifiers, cones or cyclones use the spiral air flow action or acceleration within a chamber to separate or classify solid particles. Powders suspended in air or gas enter the cyclone and the heavier particles spiral out and down where they are collected. The air and finer particles flow up to the top where they may be passed to another cyclone with finer classification capability. A cyclone is essentially a settling chamber where the effects of gravity (acceleration) have been replaced with centrifugal acceleration.

Concentrating tables or density separators screen bulk materials or minerals based on the density (specific gravity), size and shape of the particles. This group includes jigging equipment, hindered-bed settling devices, shaking table, spiral concentrators, concentrating or wet tables, hydraulic concentrating tables, constriction plate separators or specialised settling vessels. Most concentrating or density-separation equipment is hydraulic or water-based, although pneumatic or air-based systems are also available.

Electrostatic separators use preferential ionisation or charging of particles to separate conductors from dielectrics (non-conductors). The charged dielectric particles are attracted to an oppositely charged electrode and collected. The particles may be charged through contact electrification, conductive induction or high tension (ion bombardment).

Floatation systems separate hydrophobic particulates from hydrophilic particulates by passing fine air bubbles up through a solid-liquid mixture. The fine bubbles attach to and lift or float the hydrophobic particles up to where they are collected.

Magnetic separators use powerful magnetic fields to separate iron, steel, ferrosilicon or other ferromagnetic materials from non-magnetic bulk materials. The magnetic field may be generated by permanent magnets or electromagnets.

Rake, spiral and bowl classifiers use mechanical action to de-water, de-slime or separate coarse bulk materials from finer materials or liquids. Drag classifiers consist of a chain-link conveyor or endless belt that is dragged through a solid-liquid mixture. Rake classifiers lift solid-liquid mixtures up onto a plate with a screen or rake. Spiral classifiers use an Archimedes pump screw to lift solid-liquid mixtures up onto a screen for de-watering. Bowl classifiers, bowl de-silters, hydroseparators or countercurrent classifiers are other types of mechanical classifiers.

Trommels are large rotary-drum-shaped devices with a grate-like surface with large openings. Trommels are used to separate very coarse materials from bulk materials such as coarse plastics from finer aluminium recycled material, coarse inorganic materials from organic wastes or large ore chunks from finer minerals.

Water classifiers such as elutriators and classifying hydrocyclones use settling or flow in water or a liquid to separate or classify powdered materials based on particle size or shape.

3 Emerging Recycling and Recovery Technologies

The technology areas considered as being important to the future of WEEE recycling can be categorised according to their potential role in the process:

- Disassembly
- Comminution (size reduction)
- Separation
- Thermal treatment
- Hydrometallurgical extraction
- Dry capture techniques
- Biotechnological capture techniques
- Sensing technologies
- Design for recycling, and reverse supply chain technology

3.1 Automated Disassembly

Disassembly is seen by many as an essential element, even for relatively well-defined input streams, if value is to be extracted. The high labour burden of this stage is driving research into automated techniques. In automating the process the main issues to be addressed are: imaging, recognition and robotics. Demands on these are reduced by attention to upstream issues.

Disassembly can be simplified through design features that employ mechanical, chemical, thermal, electromagnetic and biological means. Biodegradable components is an interesting new area of development, while an important new class of materials is shape memory polymers (see 'Design for Recycling'). Logistics and process planning of disassembly are very active research areas. Simulation tools are being developed that allow critical assessment of disassembly system designs, and some are commonly available. RF (radio frequency) tagging is an established technology whose future potential for automated disassembly is now recognised. For example, in Japan where emphasis is placed on logistics and voucher systems, new technology integrates RFID (radio frequency identification) tag information with comprehensive databases *via* the internet.

Relevant demonstrations of robotics are restricted to a few items, such as PCBs, which have predictable shapes and components. Some researchers are now extending this to more than one product through incorporation of computers and databases in a 'hierarchical' approach that groups items into families. Some research groups are incorporating electromagnetic sensors for location of pieces and identification of materials within robotic control systems.

3.2 Comminution

Shredding, crushing, pulverising, grinding and ball milling are all relatively conventional methods for reducing particle size. These are mainly mature technologies, and there is less long-term research in this area than in the others, although use of cryogenics is being studied. Advancements in automated recycling are not likely to depend crucially on developments in this technology area.

3.3   Separation

Efficient separation is a prime requirement for effective WEEE recycling, and can reduce reliance on dismantling. A range of sorting systems is available for separating materials in general scrap, after comminution, according to properties such as weight, size, shape, density and electrical and magnetic characteristics. Some research groups are carrying out characterisation studies in order to determine costs, and to optimise the level of particle size from the comminution process. Particle size as low as 5 mm to 10 mm is preferred on technical grounds, but is more costly to generate. Although new techniques are being developed, much of the novelty in WEEE separation is expected to emerge from adapting existing techniques, and from novel combinations. Initially this is usually for specific input streams, but more sophisticated routes are being developed to handle a wider variety of input items.

Some researchers are combining separation techniques, for example, screw threads within floatation tanks, and jiggers or conveyor belts with water added, in some cases with superimposed magnetic fields. In other areas, existing techniques are being enhanced: for example, eddy current methods with rotating electrodes or with pulsed excitation, and corona separators with novel electrode designs. Electrostatic separators are good for extracting plastics, but are limited to relatively small particle size. Sand-based fluidised beds for gravity separation are also under investigation. Value may be extracted even from residual powder materials in processes under research in the automotive industry. Following pyrolysis and magnetic separation, sufficient carbon and iron dust is retrieved to render the process viable.

Separation of wire is a specific requirement. One patented method forms the wire, after crushing the other components, into coil shapes that allow gravity separation. Another approach uses spiked drums that both rotate and oscillate to facilitate separation and subsequent disengagement.

A major advance in separation is likely to come through sensing methods. Opto-electronic sorting is now being incorporated into research systems. The construction sector is developing electromagnetic field methods for both sensing and sorting. At low field levels, enhanced with pulsed excitation, signature combinations of conductivity and permeability of each material allow them to be discriminated. Higher fields eject the metal pieces from a conveyor belt. Signals are fed to a processor that activates appropriately positioned air nozzles to effect separation in the falling zone. Another system under research in the construction sector uses a similar intelligent ejection unit, but linked to a camera, and sorting on colour, shape and position. X-ray systems are also under investigation, and could equally be linked to such an automatic ejection system.

3.4   Thermal Treatments

Thermal treatments have the advantages of greatly reducing bulk and avoiding liquid effluent for the primary recycler. Pyrometallurgical routes are suitable for the recovery of metal values, *e.g.* from PCBs, which contain 29%wt of valuable metals, even though comprising only 3%wt of WEEE. There are established operators in place who incinerate the non-metallic content to produce ash, which can be used as feedstock in the pyrometallurgical processes. The final products tend to be partly refined metal ingots, such as ferrous, aluminium, mixed Pb/Sn and, most importantly, a copper-rich precious metals mix, which require further treatment by specialist refiners. Continuing research into these methods could improve the quality of metals recovered.

Encapsulation is a further thermal approach being developed. Either glass or binder can be employed to produce low-grade block products for use in construction. Harmful contents are safely sealed in, but the opportunity to recycle valuable resources is lost. Research topics include use of vacuum, thermal plasmas and lasers to enhance the thermal treatment. Other emerging variants include processes for copper-rich and iron-rich items. The latter is adapted from the car industry, but claimed to be suitable for WEEE.

3.5   Hydrometallurgical Extraction

Hydrometallurgy offers the possibility to achieve more selective metal recovery and to reach higher recycling percentage targets. Indeed these methods are often used by specialist refiners of precious metals following initial pyrometallurgical extraction. However, future research must address the need to avoid the use of hazardous materials (*e.g.* cyanide, hydrogen peroxide, fluoroboric acid) and the production of secondary waste streams. The search for still less hazardous reagents continues: thiourea and thiosulfate are candidates, as well as solvents such as polyhydric alcohol, ketones, polyether or cyclic lactone. Stability, process control, reagent recyclability, and economics are outstanding issues to be addressed in future research work. Techniques being applied in ongoing research include microwave leach enhancement, micro-biologically enhanced leaching, ionic liquid extractive technology and selective electrochemical recovery.

3.6   Dry Capture Technologies

Emerging dry capture technologies relate mainly to extraction systems and filtration, enhanced by the advent of nanotechnology, such as ultrafiltration. Little application to WEEE is found, but the generation of fine powder, which comprises 4% of WEEE arisings and contains valuable metals, may promote interest in this area in the future.

3.7   Biotechnological Capture

An interesting approach to be explored further in future research is the application of microbial cells to metal recovery from leachate solutions derived from WEEE sources. Many micro-organisms have developed an ability to capture metals as a way of dealing with their presence in the environment. The use of renewable biologically derived materials is also an interesting approach to be explored further in the future. A good example is chitosan, which is readily available as the structural element in crustaceans' exoskeletons, and can be chemically modified to improve its selectivity and capacity for metals, particularly precious metals.

3.8   Sensing Technologies

Improved sensing methods could greatly increase the effectiveness of WEEE recycling. They are crucial to implementation of automated disassembly and can facilitate great improvements in separation. Opto-electronic sorters, which use conventional imaging devices to discriminate on shape and colour, have been developed for various industries. Augmentation by electromagnetic sensing permits identification of metals, as well as of rubbers and plastics, allowing selective ejection of the identified items in automated separation processes. Laser-Induced Breakdown Spectroscopy (LIBS) is a laboratory technique that is being adapted for online operation in separation processes, and to which enhancements such as pulsing are being applied. It is useful for heavy metals, and is of particular interest because it can detect brominated flame retardants. High accuracy and operational speed have been demonstrated. Precise identification of plastics is key to economic viability of WEEE recycling. Current recycling methods generate low-grade mixed plastics, suitable only for de-rated applications. Laboratory analytical techniques that are under research for adapting to this application include laser-induced fluorescence and X-rays. Adaptations of electromagnetic sensing for conducting materials are also being developed.

3.9   Design for Recycling and Inverse Manufacturing

Disassembly is a major cost in WEEE recycling that can be greatly reduced in the future through equipment design. Efficient use of resources can also be assured. Fundamental design principles for realising this have been set out by Envirowise. Other centres promoting this approach include the Sustainable Design Network, led by Loughborough University, and SUMEEPnet at the University of Surrey. 'Inverse Manufacturing' is a complementary activity that aims to support a reverse supply-chain infrastructure for the re-use of components and sub-assemblies. A dedicated forum for this was established in Japan as long ago as 1996. Shape-memory metals and polymers, which are materials that return to a pre-determined shape on heating to a transition temperature, are a major new development that will assist disassembly. Components made from these materials can be designed to release when heated. This has been demonstrated on LCD screens and has been incorporated into some Nokia mobile phones. Microwave heating can greatly increase the speed of the process. The Japanese company Diaplex produces a range of shape-memory fasteners, while the Fraunhofer Institute is developing detachable joints based on shape-memory snaps, screws and foam supports.

Both current and emerging technologies for recovery and recycling are summarised in Table 1.

4 Printed Circuit Boards

Printed circuit boards represent one of the most interesting waste fractions from electric and electrical equipment with approximately 90% of the intrinsic value of populated (*i.e.* boards with components mounted) scrap boards being in the gold and palladium content, albeit representing <0.05% by weight of the material composition.

Methods currently employed and envisaged for future implementation utilise the intrinsic characteristics of WEEE streams, including density differences, magnetic and electrical conductivity differences, polyformity, chemical reactivity and electropositivity. The integration at varying levels of mechanical and hydrometallurgical separation with pyrometallurgical and electrochemical approaches bring together the four major recycling process methodologies to facilitate a potential total recovery-based approach for PCB waste.

4.1 Overview

Whilst information in respect of the amount of electronic scrap generated within the UK is published by ICER (Industry Council for Electronic Recycling), specific figures relating solely to printed circuit board scrap are less readily available and perceived quantifications vary greatly. It would appear from discussions with key recycling industry personnel that some 50 000 tonnes per annum of PCB scrap is currently generated within the United Kingdom, of which perhaps 40 000 tonnes per annum comprises populated boards with the remainder being either unpopulated or associated board manufacturing scrap, such as off-cuts, *etc.*

Of the 50 000 tonnes per annum of estimated PCB scrap, it is further estimated that only some 15% of such is subject to any form of recycling with the remainder being consigned to landfill. Approximately 60% of the estimated landfill demand of 42 500 tonnes per annum is believed to be consigned within the total redundant equipment package. A proportion of what would primarily be landfill demand is met by off-shore shipments to China for disassembly and pyrolysis. The recycled board waste effectively comprises only those boards having inherent value by virtue of their contained precious metal content. Recycling in the current sense is purely in respect of the recovery *via* smelting of the metal content with the vast majority of boards being refined at either:

Union Menieur (Hoboken, Belgium)

Boliden (Sweden)

or Noranda (Canada).

Table 1 Current and future technology summary.

Process stage	Process	Current technologies		Future technologies	
		Technique	Comment	Technique	Comment
Disassembly	Manual	Pre-sorted waste then physically checked			
	Impaction	Mechanical dismantling under force with chains			
	Shredding/ Fragmenting	Various shredding devices – similar to above			
	Automated	QZ machine	Combination of devices above and sepn methods below	Imaging and recognition	
				Robotics	
				Enhanced fastenings	
				RF tagging	
				Cryogenics	Used in combination with other size-reduction methods
Size reduction	Crushing	Hammer mill			
Separation	Size	Granulation Screeners/classifiers			
	Magnetic	Low-intensity drum High-intensity magnetic field			

Density (may require separation by size first)	Air tables Water tables Cyclones		
Conductivity	Eddy current Electrostatic	Rotating eddy currents Corona discharge	Develop effect on larger particles
Other characteristics	Triboelectric Flotation (hydrophobicity)	No separation	Combines recovery steps to treat the entire stream
		Carbonisation	Used on powder from shredders to recover material
		Opto-electronic sensing Heat treatment	Pyrometallurgical processes Pyrolysis
		Heat treatment Encapsulation Hydrometallurgical Dry capture Biotechnological capture Sensing	

Within the UK, both Johnson Matthey and Engelhard accept scrap PCBs through their smelters but the costs are such that only boards with very high precious-metal content are processed. Boards shipped for smelting have invariably been subject to 'upgrading' *via* shredding and magnetic and additional classification.

4.2   Recycling

Populated printed circuit board assemblies may be anticipated as having the following approximate material composition:

GRP (glass-reinforced plastic)	>70%
Copper	16%
Solder	4%
Iron, ferrite (from transformer cores)	3%
Nickel	2%
Silver	0.05%
Gold	0.03%
Palladium	0.01%
Other (bismuth, antimony, tantalum, *etc.*)	<0.01%

General routes which may be followed for recycling would comprise:

- component recycling *via* disassembly
- materials recycling *via* mechanical processing, pyrometallurgy, hydrometallurgy or a combination of these techniques

Scrap PCBs forwarded to a smelter are rarely subject to any form of upgrading other than selective disassembly, grading and shredding to reduce bulk volume, due to the inherent loss in precious-metal content that is perceived to occur within additional separation/classification processes. It is not uncommon for companies engaged in general waste recycling of PCBs to have observed precious-metal content loss of *c.* 10% even with wet mechanical separation processes. With dry mechanical separation processes the potential loss may be anticipated as being somewhat higher than this and levels of up to 35% loss have been reported in some instances of boards having high PMG (precious metal group)-bearing component populations. This is considered in the main to be a function of the nature of the interface at which the precious metal is present within populated components and adherence of such to comminuted plastic particulates.

Specialist markets exist in the US and Europe for the recovery and refurbishment of components from PCBs for sale in the second-user marketplace and whilst robotic technologies operating from maintained populated board databases offer the capability for cost-effective component identification and disassembly, there are many experts in the field who consider that the low cost and high technology of new components will prove to place serious limits on the long-term viability of component recycling. There continue to be, however, emerging techniques for disassembly which embrace thermal methods and in concert with solder removal *via*, for example, the deployment of shearing rollers.

Pyrolytic treatment normally comprises the ignition and melting of ground feedstock within a furnace at temperatures of *c.* 1200 °C *via* air injection and, whilst a small amount of oil is normally required, much of the energy is provided by the organic components of the scrap. The organic constituents of scrap boards are destroyed at such temperatures and toxic emissions from such destruction are addressed *via* afterburners in the off-gas ducting operating at 1200–1400 °C. The metal produced is called 'black metal' and is generally a copper-rich product which is subject to electrorefining, with the precious metals being ultimately recovered from the anodic sludge *via* a leaching, melting and precipitative route.

Whilst the vast majority of scrap or redundant printed circuit board assemblies which currently enter the recycling route, primarily for their precious metal content, are subject to pyrolytic treatment (smelting) *via* initial primary mechanical treatment, there are a number of enhanced mechanical treatment approaches, either commercialised or in the course of being commercialised, which seek to add value prior to pyrolysis and to generate a separated plastic component to effect true recycling. In Germany for example, FUBA have commercialised the generation of a 92–95% metal stream output from scrap unpopulated PCBs *via* a mechanical process route involving shredding, granulation, magnetic separation, classification and electrostatic separation. Plastic stream output from this facility has found application in extrusion casting in the instance of a fibreglass-rich component and as a filler in building materials in the case of a powder-generated fraction. These downstream applications for separated plastic fractions have recently been superseded by FUBA's own development of their combined use in chemical-resistant plastic pallets, which represents both a higher added-value application and one that overcomes market restrictions and cost barriers in the commercial supply of additive materials. Commercial mechanical recycling systems are also being currently offered as turnkey plants deploying comminution, magnetic and eddy-current separation (for ferrous and aluminium fractions), classification, electrostatic separation and secondary treatment to generate metallic fractions, non-conductive and ferrous fractions from scrap PCB assemblies.

Although scrap laminate materials, in the form of offcuts, *etc.*, are more accurately defined as associated PCB waste and may be subject to pyrolysis for both ultimate copper recovery or the generation of a copper ash for application in fertiliser production, hydrometallurgical approaches have been commercialised in the US for the treatment of such *via* dissolution in sulfuric/nitric acid leachants and subsequent electrolytic copper recovery. In addition to processing scrap unpopulated PCBs, FUBA's German facility also processes materials such as laminate offcuts.

4.3   Current Disposal Hierarchy

The primary sources of scrap PCBs are from OEMs (Original Equipment Manufacturers), PCB manufacturers, end users (corporate or individual) and equipment dismantlers. Output from these sources is either directly to recyclers or specialist recovery operations or indirectly to recyclers *via* disposal contractors. Component disassembly may be effected either by the OEMs for resale or re-use within the supply chain or by recyclers and is invariably undertaken manually. The result of manual disassembly is that the cost impact of such renders component recovery viable only in instances of relatively high-value elements or in the case of recyclers where additionally the component presence, such as transformer cores, may either detract from the final residual value at a smelter or hinder any shredding/granulation processes.

The larger recyclers will generally effect a level of disassembly of scrap PCBs followed by sorting, grading and shredding operations with added value to the final ground product possibly being effected *via* removal of iron and aluminium content by the deployment of magnetic and eddy-current separation. The output from the recycler will be either to landfill or to a smelter, and the only boards being forwarded for smelting and subsequent recovery of the metallic constituents are those which have been graded as containing gold/precious-metal content to make such economically viable. All non-precious-metal-bearing board scrap is consigned to landfill.

The input to landfill currently represents some 85% of all the PCB scrap-board waste generated and is generally a combination of that arriving from recyclers, disposal contractors or specialist recovery operations with some 60% of the scrap being consigned to landfill within its original equipment.

It is estimated that some 30 companies within the UK are actively engaged in the handling of scrap PCBs, with approximately 15 of these supplying the input directly to smelters.

Scrap PCBs are generally subject to grading into three categories which essentially mirror their inherent precious-metal content. These are referred to as H (high)-, M (medium)- and L (low)-grade scrap. Low-grade material would comprise television boards and power supply units having heavy ferrite transformers and large aluminium heat sink assemblies; laminate offcuts would also be considered as low-grade material. Medium-grade scrap would be that from high-reliability equipment with precious-metal content from pin and edge connectors and with little incumbent material such as aluminium capacitors, *etc.* High-grade material would comprise discrete components, gold ICs, opto-electric devices, high-precious-metal-content boards, gold-pin boards, palladium-pin boards, thermally coupled modules from mainframes, *etc.* These gradings represent essentially inherent precious-metal content and even the low-grade material will tend to have a very small PMG content – it is possible to effect upgrading from a low to medium category *via* selective manual disassembly of high percentage mass ferrous and aluminium components.

Within the UK a small amount of scrap PCBs do find their way through what may be termed specialist recycling operations albeit, with the notable exception of FUBA in Germany which is taking some UK unpopulated scrap boards, such tend to be limited to operations concerned solely with precious-metal recovery. It is estimated that, of the total amount of board scrap generated, <1% finds its way to such specialised recycling operations.

As >70% of the mass of boards forwarded to smelting comprises GRP (glass-reinforced plastic) which is destroyed pyrolytically, it is thus clear that of the estimated 50 000 tonnes per annum of UK-generated scrap that some 15% of the metals content is recovered with *c*. 95% of all the scrap being either consigned to landfill or subject to thermal destruction.

It should be noted that there is an increasing trade in the export of scrap PCBs offshore into mainland China for 'recycling'. Within Europe, between 1000 euros and 5000 euros per tonne are currently being offered for PCB scrap on a three-level grading basis. It is apparent from recent studies that the methodology pursued within China would embrace the use of cheap labour for disassembly in a non-controlled manner with subsequent smelting of the depopulated board assemblies. The European Union is, of course, a signatory to the Basel Convention which has sought to adopt a total ban on the export of all hazardous waste from rich to poor countries for any reason, including recycling. PCB scrap does, however, fall largely outside the definitions of hazardous waste which gives a limit of lead at 3% as a threshold. On the assumption that other specifically noted hazardous elements such as mercury or cadmium are absent, the level of lead within populated PCB waste is generally in the order of 2% and it is considered that the majority of scrap PCB would fall outside the restrictions imposed by the Basel Convention. In respect of defined hazardous wastes the Basel Convention additionally calls on all countries to reduce their exports of such to a minimum and, to the extent possible, deal with their waste problems within national borders. Indeed, this is an obligation of the Basel Convention regardless of the level of waste management technology in the importing country.

4.4 Economics of Recycling

To help appreciate the economic drivers involved in the recycling of scrap PCBs, Table 2 provides a tabulation of the approximate intrinsic value of typical medium-graded populated circuit board waste: (metal values are based on June 2002 LME (London Metal Exchange) levels).

The major points which may be seen from this breakdown and related cost factors are:

- Approximately 90% of the intrinsic value of the scrap boards is in the gold and palladium content.
- Commercial smelter operations would typically credit between 92% and 98% of the sampled precious-metal value.
- Basic charge levied by a commercial smelter would be in the order of £400–£1000 per tonne inclusive of sampling and shipping costs.

Table 2 Typical intrinsic values of scrap populated PCB waste.

Component	% by wt	Value per kg	Intrinsic value (£/kg)	Intrinsic Value (%)	% of value from smelter
Gold	0.025	6500	1.63	59.4	98
Palladium	0.010	8000	0.8	29.2	92
Silver	0.1	70	0.07		95
Copper	16	0.8	0.13		96
Tin	3	3	0.01		
Lead	2	0.3	–		
Nickel	1	5	0.05		
Aluminium	5	0.9	0.05		
Iron	5	0.1	–		
Zinc	1	0.8	–		
TOTAL			2.74		

It is clear that, for scrap PCBs containing less than such levels of precious metals and which would be classified as low grade, it would be uneconomic to process *via* smelting. It is equally clear, bearing in mind that recyclers have to purchase scrap PCB assemblies, that maximum yield of contained precious metals is realised and that currently such is best attained *via* shredding of boards without additional comminution and classification to reduce bulk volume.

4.5 Future Developments

The cost effectiveness of pyrolytic recycling for scrap-board assemblies, apart from those with relatively high precious-metal content, coupled with increasing ecological concerns, has cast doubt over the long-term viability of this methodology and has provided impetus for developments of more sustainable approaches embracing both mechanical and hydrometallurgical technologies.

To effect significant increases in the levels of recycling being undertaken it is clear that any approaches must embrace, in a cost-effective manner, the treatment of scrap-board assemblies other than those bearing precious-metal content. Newer approaches should also seek to more realistically address the issues of total recycling with recovery and downstream applications developed for the GRP element of scrap boards which represents in excess of 70% of the total input.

All existing and potential treatment approaches involve mechanical or mechanical/hydrometallurgical methodology. Due to the nature of the input material base even the most sophisticated hydrometallurgical treatment models involve a level of mechanical treatment. Mechanical treatment systems have currently realised a far greater level of development than hydrometallurgical ones and whilst the output from such invariably finds final treatment pyrolytically within a commercial smelter, there are many distinct advantages in the operation of enhanced mechanical-treatment systems. The major advantages of mechanical systems lie, of course, in their basically 'dry' mode of operation without the use of any operational chemistry as would be necessitated with a hydrometallurgical-based system. The use of any chemical approach will create a downstream environmental demand, from either liquid or gaseous pollution which must be addressed, and further should be addressed deploying a developed sustainable approach which does not itself create an off-site disposal demand from secondary waste. It should be noted, however, that hydrometallurgical approaches do offer a genuine treatment alternative to smelting and the possibility of realising somewhat higher metal-recovery yields. This latter point is of great significance when dealing with high-value scrap PCBs and the inherent loss from process of precious metals which may be evident within a mechanical route involving comminution, separation and classification of all materials. Ultimately, of course, the issues of cost-effectiveness and ecological concerns must be addressed, and both of the stated approaches may be fundamentally improved by being deployed on input material which is more primarily metallic and has maximised the prior removal of plastic fractions for downstream applications. It would appear that the constraints on total recycling would even be far better addressed *via* the development of hydrometallurgical approaches for high-value board scrap whilst utilising a total mechanical approach for low-grade scrap.

4.6   Characteristics of PCB Scrap

PCB scrap is characterised by significant heterogeneity and relatively high complexity, albeit with the levels of complexity being somewhat greater for populated scrap boards. As has been seen in respect of materials composition, the levels of inorganics, in particular, are diverse, with relatively low levels of precious metals being present as deposited coatings of various thicknesses in conjunction with copper, solders, various alloy compositions, non-ferrous and ferrous metals. In spite of the inherent heterogeneity and complexity, there are differences in the intrinsic physical and chemical properties of a broad spectrum of the materials and components present in scrap PCBs, and indeed electronic scrap as a whole, to permit recycling approaches in separating such into their individual fractions. The following characteristics ultimately govern mechanical and hydrometallurgical separation and what it is based upon, such that current and potential recycling techniques and infrastructures have been envisaged, developed and implemented.

Density Differences
The differences in densities of the spectrum of materials contained within scrap PCBs has formed the basis of separation based on such subsequent to their liberation as free constituents. The specific-gravity ranges of typical contained materials are as shown in Table 3.

With these densities not being significantly affected by the addition of alloying agents or other additives, it is predictable that the deployment of

Table 3 Specific gravity ranges of materials within PCBs

Materials	Specific gravity range (g/cm³)
Gold, platinum group, tungsten	19.3–21.4
Lead, silver, molybdenum	10.2–11.3
Magnesium, aluminium, titanium	1.7–4.5
Copper, nickel, iron, zinc	7–9
GRP	1.8–2

various density-separation systems available within the raw materials process industry may be utilised to effect separation of liberated constituents of a similar size range. The utilisation of density differences for the recovery of metals from PCB scrap has been investigated on many occasions and air classifiers have been used extensively to separate the non-metallic (GRP) constituents, whilst sink-float and table-separation techniques have been utilised to generate non-ferrous metal fractions. Air-table techniques which effectively combine the actions of a fluidised bed, a shaking table and an air classifier, have been successfully implemented in applications involving a diversity of electronic-scrap separations. It is, of course, essential, as has been noted, that the feed material must be of a narrow size-range to guarantee effective stratification and separation.

Magnetic and Electrical Conductivity Differences
Ferrous materials may readily be separated with the application of low-intensity magnetic separators, which have been well developed in the minerals processing industry.

Many non-ferrous materials in respect of their high electrical conductivity may be separated by means of electrostatic and eddy-current separators. Eddy-current separation has been developed within the recycling industry since strong permanent magnets, such as iron-boron-neodymium, became available. Rotating belt-type eddy-current separation is the most extensively used approach for the recovery of non-ferrous metal fractions. In application, the alternating magnetic fields resulting from the rapidly rotating wheel mounted with alternating pole permanent magnets result in the generation of eddy currents in non-ferrous metal conductors which, in turn, generate a magnetic field that repels the original magnetic field. The resultant force, arising from the repulsive force and the gravitational force, permits their separation from non-conducting materials.

Polyformity
One of the important aspects of both PCB and electronic scrap is the polyformity of the various materials and components and the effect such can have on materials liberation. It is essential that shredding and separation processes must account for this. In eddy-current separation, the shape of conducting

components, in addition to their particle sizes and conductivity/density ratios, has a significant effect on the repulsive forces generated, which ultimately govern the separation efficiency. For instance, multiple induced-current loops may be established in conductors with irregular shapes with the induced magnetic fields counteracting each other and reducing the net repulsive force.

Liberation Size

The degree of liberation of materials upon shredding and comminution is crucial to the efficiency and effectiveness of any subsequent separation process in respect of yield, quality of recovered material and energy consumption of the process. This is especially critical in mechanical-separation approaches. The comminution of scrap PCBs has been shown to generate a high level of material liberation and levels as high as 96–99% have been reported for metallic liberation after comminution to sub-5 mm particulates. It must be borne in mind, however, that a noted and continual observation from recyclers is that liberation levels such as these are somewhat untypical of actual yields and that a fundamental constraint on mechanical processing, as has been noted, is the loss, particularly of precious-metal content, that appears to be inherent due primarily to the nature of many plastic/metal interfaces.

Chemical Reactivity

Hydrometallurgical approaches depend on selective and non-selective dissolution to realise a complete solubilisation of all the contained metallic fractions within scrap PCBs. Whilst all hydrometallurgical approaches clearly benefit from prior comminution, such is primarily undertaken to reduce bulk volume and to expose a greater surface area of contained metals to the etching chemistry. Selective dissolution approaches may utilise high-capacity etching chemistries based on cupric chloride or ammonium sulfate for copper removal, nitric acid-based chemistries for solder dissolution and aqua regia for precious metals dissolution, whilst non-selective dissolution may be carried out with either aqua regia or chlorine-based chemistry.

Electropositivity

Dissolved metals generated *via* chemical dissolution are present as ionised species within aqueous media and may be recovered *via* high-efficiency electrolytic recovery systems. In the instance of selective dissolution, a single metal is recovered as pure electrolytic grade material, usually in sheet form, from the spent etching solution, with certain etching chemistries permitting regeneration of the liquors for re-use as etch chemicals. In the instance of selective dissolution, use may be made of the differing electropositivities of the contained ionised metallic species for selective recovery of metals at discrete levels of applied voltage.

4.7 Emerging Technologies

4.7.1 Mechanical Approaches

As may be anticipated, all of the work undertaken on mechanical systems has been with the primary objective of enhancing separation yield of the various fractions, particularly the precious-metal-bearing ones. The basic mechanical techniques deployed in the treatment of scrap PCBs and electronic assemblies have been adapted or adopted from the raw-materials processing sector and refinement has sought to address both yield constraints and ultimately cost-effectiveness of the approaches, either used singly or in an integrated manner. The problems associated with yield were apparent from early attempts to produce a model methodology for handling all types of electronic scrap, as instanced by the US Bureau of Mines (USBM) approach in the late 1970s and early 1980s. The separation route, developed up to a 250 kg per hour pilot plant, comprised shredding, air separation and magnetic, eddy-current and electrostatic separation to generate aluminium-rich, copper-rich (including a major precious-metal fraction), light air-classified and ferrous fractions. The yield, however, was such that no commercial uptake of this approach has been instanced. The relatively poor yields or levels of separation obtained from this approach were undoubtedly due to the use of a standard hammer mill having no provision or levels of refinement to cope with clear comminution of aluminium, the use of a ramp-type eddy-current separator of low capacity and selectivity and the use of a high-tension separator for metals/non-metals which has since been demonstrated as having low capacity and high susceptibility to humidity.

There was little further meaningful development work on the implementation of mechanical treatment approaches until the early 1990s, when Scandinavian Recycling AB in Sweden implemented their mechanical concept for electronic scrap handling. This did not specifically address the treatment of scrap PCBs but rather removed PCBs for specialist treatment as part of the pre-sorting stage. Subsequent to this, development work in both Germany and Switzerland has seen the implementation of mechanically based approaches for the handling and separation of electronic scrap, with the work at FUBA in Germany dedicated to scrap PCBs being a notable example of this activity.

In 1996, Noell Abfall and Energietechnik GmbH in Germany implemented a 21 000 tonnes per annum plant with the capability of handling a wide variety of electronics scrap, but specifically intended for redundant-telecommunications scrap. The system again involves PCB scrap and the inherent precious-metal content being subject to prior manual disassembly and the overall methodology deploys a three-stage liberation and sequential separation routes with ferro-magnetics removal *via* overhead permanent magnets and eddy-current techniques, due their ability to optimise handling of fractions in the 5–200 mm particle-size range. Air-table techniques were utilised for the separation of particulate fractions in the 5–10 mm, 2–5 mm and <2 mm ranges, respectively.

Mechanical and physico-mechanical approaches to the treatment of scrap PCBs may be deployed as stand-alone treatment stages, *i.e.* pulverisation, magnetic separation or integrated into a complete treatment system, with the output being metallic and non-metallic fractions. The metallic output would be destined for pyrometallurgical refinement *via* smelting whilst the non-metallic output would find applications in the secondary plastics marketplace, or be utilised within dedicated developed applications. As has been noted, FUBA have developed their total mechanical treatment system, albeit only currently utilised for non-populated board scrap or ancillary laminate waste, through this latter route.

There are commercially available turnkey mechanical systems currently available for the treatment of a wide range of electronic-scrap materials, including both populated and non-populated PCBs. One such is that developed by Hamos GmbH in Germany which is an automated integrated mechanical system comprising the following stages:

- Primary coarse-size reduction – this is accomplished with a shredder having multi-use rotational knives.
- Coarse ferrous-metal separation – accomplished with rare-earth magnets sited above an oscillating conveyor belt feed to allow high-efficiency ferrous separation across a range of particle sizes.
- Pulverisation – circuit board assemblies are pulverised within a hammer mill utilising high-abrasion-resistance hammers and liners and proprietary grates, with the action of the mill inducing a 'spherising' effect on the metallic particulates.
- Classification – utilising self-cleaning sieves.
- Electrostatic separation – virtually complete separation of metallic fractions with recirculation of mid-range particulate fractions.
- Further size reduction – secondary pulverisation to effect size reduction on oversized particulates.

The Hamos system additionally can incorporate density separation for aluminium extraction and dust-generation treatment of any such outfall from the hammer mills *via* secondary electrostatic separators. The complete conveyorised-based systems are operated at negative pressures to eliminate any airborne pollution and are currently available with treatment capabilities up to 4 tonnes per hour of input feed. All product from the system, *viz.* mixed plastic, metallic and extracted ferrous and aluminium, is bagged automatically for onward shipment.

Considerable work has been undertaken on enhancing the effectiveness of mechanical treatment systems, such as the development of newer pulverising process technology *via* the application of multiple pulverising rotors and ceramic-coated systems. This has enabled the generation of sub-mm particulate comminution which has, in turn, enabled the efficiency of subsequent centrifugal separation techniques to realise 97% copper-recovery yields. The effectiveness of the pulverising process has been improved by the adoption of dual pulverising stages: a crushing process and a fine pulverising process. The crushing process combines cutting and shearing forces and the fine pulverising process combines shearing and impact forces. This approach is shown in Figure 1. With such effective particulate comminution, both screen separation and gravity separation have been investigated and conclusions drawn that the most effective approach was by gravity, using a centrifugal classifier with a high air-vortex system.[1]

Researchers at Daimler-Benz in Ulm, Germany, have developed a mechanical treatment approach which has the capability to increase metal-separation efficiencies, even from fine dust residues generated post-particulate comminution, in the treatment of scrap PCB assemblies. They considered a purely mechanical approach to be the most cost-effective methodology and a major objective of their work was to increase the degree of purity of the recovered metals, such that minimal pollutant emissions would be encountered during subsequent smelting. Their process comprises the initial coarse size-reduction to *c.* 2×2 cm-dimensioned fractions, followed by magnetic separation for ferrous elements, which is followed by a low-temperature grinding stage. The embrittlement of plastics components at temperatures $<70\,^{\circ}\mathrm{C}$ was found to enable enhanced separation from non-ferrous metallic components when subjected to grinding within a hammer mill. In operation, the hammer mill was fed

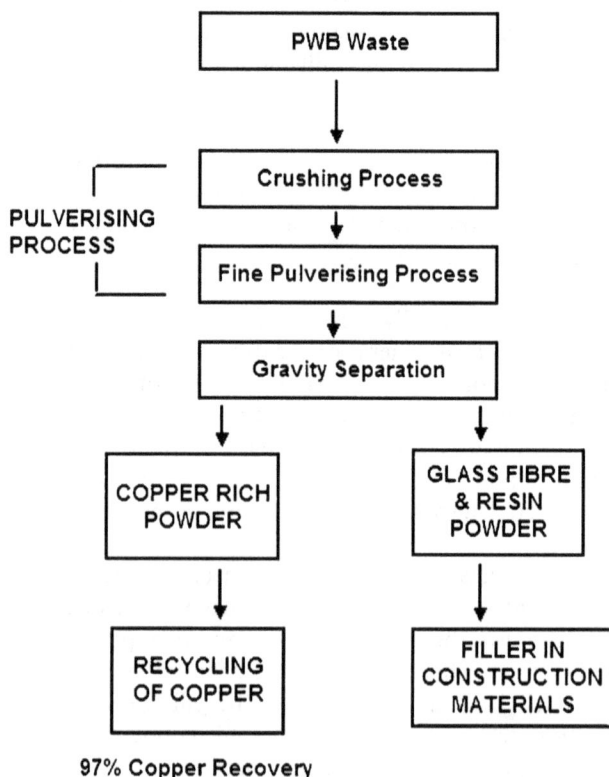

Figure 1   Integration of pulverising within scrap PCB recycling.

with liquid nitrogen at −196 °C, which served both to impart brittleness to the plastic feedstock constituent and to effect process cooling; additionally, the grinding of material within such an inert atmosphere eliminated any likelihood of oxidative by-product formation from the plastics, such as dioxins and furans. Subsequent to this enhanced grinding stage, the metallic and non-metallic fractions were separated *via* sieving and electrostatic stages. Cost analyses undertaken by Daimler-Benz engineers have indicated that such a process may be economically viable even when dealing with relatively low-grade PCB scrap having little precious-metal content. Ongoing activities are concerned with development of the treatment of separated plastic fractions in conjunction with Mitsubishi Heavy Industries who have set up a gasification and methanolysis plant to such effect.[2]

Air-table separation systems have been researched with a view to effecting separation of metallic and plastic components from an input feed of screened 7 mm shredded particulate scrap PCBs post ferromagnetic separation.[3] Recovery rates for copper, gold and silver of 76, 83 and 91%, respectively, were considered to validate the approach, but only for low-grade PCB scrap or general electronic scrap.

4.7.2 Hydrometallurgical Approaches

A number of hydrometallurgical approaches have been developed through to pilot-plant stage with preliminary cost studies indicating the potential recovery of all materials, with the exception of discrete components, at an operational profit of some $200 per tonne.

In the USA, a methodology based on solvolysis has been developed to enable both the more efficient recovery of metals and the recovery of plastic materials, such as epoxies, at high quality and with the additional benefit of having the capability to extract both halogens and brominated hydrocarbon derivatives.[4]

On a relatively small scale, there have been a number of hydrometallurgical approaches traditionally pursued in the recovery specifically of gold from pins and edge connectors. Such methodologies have usually been deployed on discrete edge connectors and gold-coated assemblies, which have been manually separated from the scrap board *via* the use of air knives, *etc*. The approaches have been either liberation of gold as metal flake *via* acidic dissolution of the copper substrates, or dissolution of the gold in cyanide or thiourea-based leachants, followed by electro-winning or chemical displacement/precipitation with reagents such as powdered zinc.

The use of non-selective leachants to dissolve other than precious-metal content of scrap PCBs has also received attention, and various studies have been undertaken on the viability of utilising dilute mineral acids in conjunction with subsequent metal-recovery techniques based on concentration and separation, such as solvent extraction, ion exchange, adsorption and cementation.[5]

In the UK, there have been two potentially significant development projects undertaken on hydrometallurgical approaches to the recycling of scrap PCBs, with both having demonstrated viability to a pre-pilot-plant stage. The first of these approaches is with a Cambridge-University-led consortium, which deploys a selective dissolution/electrolytic recovery route for discrete metal constituents. The solder recovery stage employs a solder-selective (non-copper etch) regenerable leachant based on fluoboric acid, which may or may not be deployed prior to mechanical pre-treatment, from which the dissolved solder may be electrolytically recovered in pure metallic form. Subsequent selective leaching of copper and PMG metals may then be carried out. The ability to selectively remove solder prior to mechanical comminution has specific advantages in enabling disassembly and component integrity and recovery. Mechanical pre-treatment methodologies followed by the Cambridge group have been shredding, magnetic separation, eddy-current separation and classification.

The second development is that of an Imperial College, London, consortium which has taken shredded and classified sub-4-mm PCB/populated PCB scrap through a single leachate route comprising electro-generated chlorine in an acidic aqueous solution of high chloride-ion activity. This has produced a multi-metal leachate electrolyte, containing all of the available metal content, at generally mass-transport controlled rates with respect to dissolved chlorine. The viability of subsequent metal recovery *via* electrolytic membrane cells with discrete metal separation has also been demonstrated.

4.7.3   Disassembly of PCBs

Disassembly is considered an integral element of realising intrinsic value on a scrap PCB assembly. As such, it is carried out at a number of levels: by the OEM or equipment manufacturer themselves for recovery of components from faulty products or over-capacity manufacture for re-use or replacement, by a specialist contractor performing this function for the manufacturer or by a recycler or disassembler for resale in the secondary component market. Such disassembly operations are practically all carried out manually, which in itself places limits on the operation in respect of the costs involved. Disassembly is considered to be an area of increasing significance in a marketplace of low-cost components and the necessity to address such in an automated low-cost manner. Disassembly may also be considered to have impact upon overall future recycling strategies – it has already been noted that the limitations of purely mechanical process routes are effectively concerned with precious-metal loss from components structures on populated boards due to the nature of the metal/non-metal interface and an effective automated-disassembly methodology could well expand the potential for mechanical turnkey approaches for all grades of scrap PCBs.

In Austria, SAT have developed automated component-disassembly methodology for the dismantling of components from scrap, redundant or malfunctioning printed circuit board assemblies. Whilst the existing production facility which has been set up deals with the recovery of relatively expensive components from faulty products and over-capacity manufacture from a number of German, Hungarian and Austrian OEMs, the potential exists to expand the application of this technology to complete component disassembly. SAT concur that the dismantling of components by any manual approach will be both time and cost intensive and have little future applicability within the overall treatment of scrap PCBs, which SAT currently estimate quantitatively as 400 000 tonnes per annum within Europe. SAT's technology essentially comprises automated component scanning and dual-beam laser desoldering, with vacuum removal of selected components. The component disassembly operation comprises the following stages:

- Scanning – read all component identification data
- Read stored component database – component-cost data stored
- Are the identified components soldered or surface-mounted?
- If mounted, disassembly *via* robot in 3–5 s (cost = 0.5 euro)
- If soldered – three types – highest quality *via* laser with minimum thermal input (18–20 s per component) – lower quality and BGA (ball grid arrays) *via* infrared heat input.

In concert with their work on mechanical treatment of scrap PCBs, the NEC Group in Japan[6] have sought to address the automation of disassembly *via* a mechanical approach. This approach is schematically depicted in Figure 2.

Equipment has been developed to remove components in a conveyorised mode *via* heating with infrared and shearing and, as a separate development, having a higher throughput rate *via* crushing with impacting rollers. Whilst both of these approaches leave the bare board intact, the former results in removal of both surface-mount and soldered components without loss of integrity. The NEC team additionally extended the heat-impacting equipment to effect residual (*c.* 4%) solder removal *via* automatic belt sanding. A clear objective of this work was to reduce the intrinsic material loss from mechanical treatment and utilise more fully the uneven material distribution between the bare boards and components.

The increasing rate and levels of redundancy of PCBs is a function of that being realised for all electrical and electronic equipment. There are some significant differences, however, in that the greatest intrinsic material value, specifically precious metals, within scrap equipment invariably is in the contained PCBs. This has led to the development of a commercial infrastructure based on the dedicated collection of PCBs and subsequent grading of such, with those having sufficient precious-metal content to justify recovery being processed within a smelter. The vast majority of scrap PCBs (*c.* 85%) are consigned to landfill either directly or within their original equipment. This represents a non-sustainable loss of finite materials resources and is placing a dramatically increasing burden on landfill.

The solution to the problem of discarded PCBs, as indeed with discarded electronic goods, is recycling, which will reduce the landfill disposal demand and encourage recovery of valuable materials and the re-use of components. A total recycling approach will provide a substantial source of ferrous, non-ferrous and precious metals together with non-metallic plastic materials. The

```
┌─────────────────────────────┐
│    SCRAP POPULATED PCB       │
└─────────────────────────────┘
              │
              ▼
      ┌─────────────────┐            ┌──────────────┐
      │  DISASSEMBLY    │───────────▶│  COMPONENTS  │
      └─────────────────┘            └──────────────┘
              │   heat + external force
              ▼
      ┌─────────────────┐            ┌──────────────┐
      │  SOLDER REMOVAL │───────────▶│    SOLDER    │
      └─────────────────┘            └──────────────┘
              │   heat, impacting force,
              │   surface abrasion
              ▼
  ┌───────────────────────────────┐
  │        PULVERISATION          │
  └───────────────────────────────┘
              │   crushing, fine
              │   pulverisation
              ▼
      ┌─────────────────┐
      │   SEPARATION    │
      └─────────────────┘
          gravity + electrostatic
         │                    │
         ▼                    ▼
┌──────────────────┐  ┌──────────────────┐
│  COPPER RICH     │  │  GLASS FIBRE &   │
│  POWDER          │  │  RESIN POWDER    │
└──────────────────┘  └──────────────────┘
```

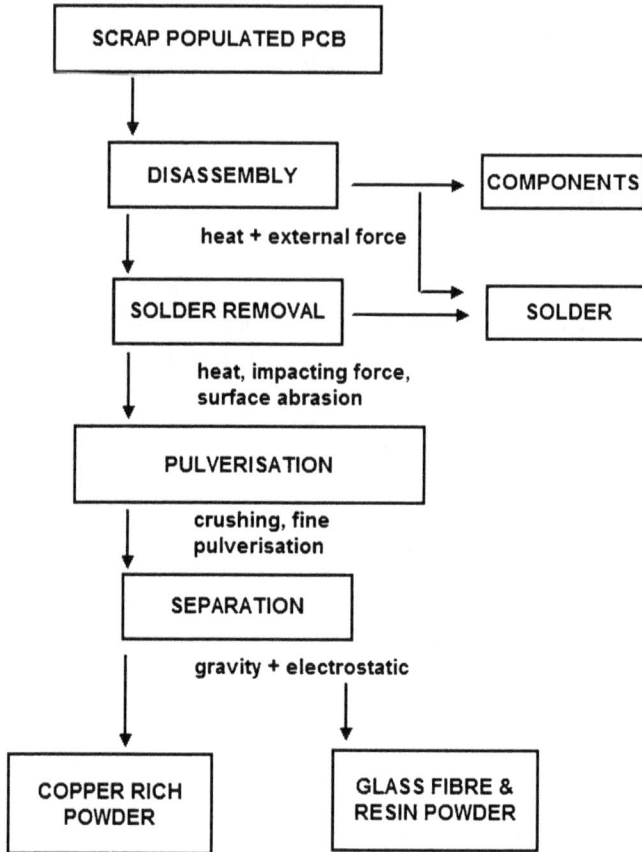

Figure 2   Integration of disassembly by mechanical means into treatment of scrap
PCBs.

necessity of focusing on new and viable recycling approaches has been
acknowledged and addressed within the European EUREKA project (EU
1140) 'A Comprehensive Approach for the Recycling of Electronics (CARE)
'VISION 2000', which was initiated to enhance the value of the recycling of
electronics by developing methods for disassembly, materials separation, and
identification and recovery of marketable products. The driving forces behind
the EUREKA project were the high value of many parts in electronic scrap and
the difficulty and inappropriateness of landfill as a disposal option. Whilst the
intrinsic value of electronic components has decreased dramatically over the
past five years, there is undoubtedly still a significant market for many recovered components and the landfill disposal option has taken on ever-increasing
significance since the EUREKA project was launched (1994).

   In terms of the products obtained from scrap PCBs, there may be considered
two recycling categories, component recycling and materials recycling, whereas
in terms of recycling techniques five categories have been noted. Various

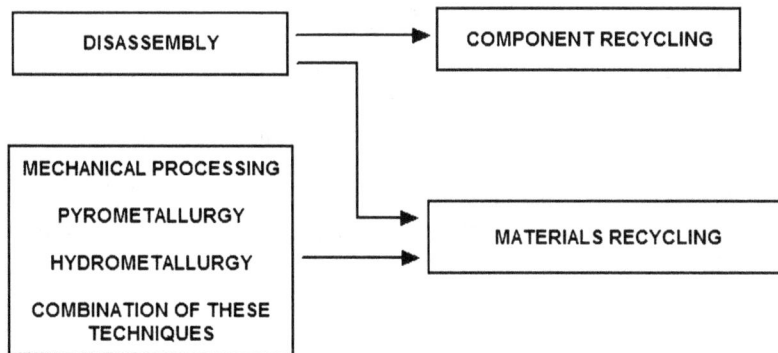

Figure 3 Recycling approaches to scrap PCBs.

recycling approaches have embraced some or all of these categories and techniques, the relationship between which is shown in Figure 3.

As has been seen, it is not uncommon for the disassembly process to be employed to segregate components and/or materials that are re-usable, identifiable or hazardous in such a manner as to maximise economic return and to minimise environmental demand, enabling subsequent processes to be performed more effectively and efficiently.

From the early attempts and approaches towards recycling of PCB scrap which have been noted in this study, together with historic work on total mechanically based routes, mechanical recycling techniques have been enhanced and evolved through to commercial implementation in Germany (FUBA). The deployment of such enhanced mechanical-separation and treatment techniques, either as a dedicated recycling approach or in concert with hydrometallurgical and pyrolytic methodologies, does provide the basis of a recycling approach to address both the level of redundant board scrap currently consigned to landfill or off-shore and the diminishing level of boards with high precious-metal content levels. It is considered highly unlikely that a single universal approach will be evolved but rather a flexibility of treatment options based on hydrometallurgy, mechanical and pyrolytic technologies, integrated with higher levels of automated disassembly. Such flexibility is considered necessary to address the variability of input material and the related variable intrinsic-content value. The necessity to develop real downstream applications for liberated plastic waste is an issue which must be addressed within a truly sustainable recycling scenario and the efforts of FUBA in Germany have demonstrated what is possible in this respect.

The studies undertaken to date have confirmed both the level of circuit-board scrap being generated within the UK marketplace and the current destination of such, with only those elements of scrap having precious-metal content following an essentially pyrolytic/smelting recycling route. The cost-effectiveness of such a pyrolytic approach for all but high-value scrap boards (currently some 15% of the total and declining with current reduced precious-metal content of assemblies), together with increasing ecological concerns, has led to the commercialisation of enhanced mechanical methodologies and the development of a number of hydrometallurgical approaches. It would appear to be a very real possibility that the integration of these second-generation mechanical and hydrometallurgical treatment routes will offer a cost-effective and more sustainable alternative methodology to pyrolysis and one which will permit the viable recycling of scrap board assemblies which are currently consigned to landfill.

Points of significance relating to the treatment of PCB scrap may be summarised as follows:

- Some 50 000 tonnes per annum of PCB scrap is generated within the UK, comprising *c.* 40 000 tonnes per annum of populated boards with the remainder being unpopulated boards and associated board waste such as laminate off-cuts, *etc.*
- Approximately 15% of this total level of scrap is subject to any form of recycling, with the balance being consigned to landfill or off-shore.
- The scrap subjected to recycling is only that which contains a relatively high proportion of precious metals (usually gold and palladium) and thus is economically viable.
- All of the scrap subjected to such recycling is treated pyrolytically within a smelter.
- >90% of the intrinsic material value of boards which may be classified as medium-grade scrap is in the gold and palladium content.
- Mechanical upgrading other than disassembly, grading and shredding for bulk volume reduction prior to pyrolysis is not undertaken due to inherent yield loss, particularly of precious metals. This loss may be typically in the order of 10% but may be much higher.
- Yield problems with mechanical treatment methodologies are more a function of the plastic/metal interface on components. For unpopulated or depopulated boards this is less of a constraint and successful commercial total mechanical recycling has been implemented for such assemblies (FUBA, Germany).
- Disassembly has traditionally been undertaken manually, but newly developed automated systems will impact upon future recycling strategies both to maximise cost-effectiveness for low-value component recovery and as an initial stage for recycling approaches to maximise yield of residual intrinsic-material value.
- Hydrometallurgical approaches offer the opportunity to eliminate metal-yield loss from recycling processes, but have potentially more significant environmental impact in implementation.
- Mechanical treatment approaches would appear to offer significant environmental and operational benefit and this is reflected in the amount of development work undertaken on such during the past 20 years, with the focus having been on improving yield and efficiency.
- Mechanical and hydrometallurgical recycling approaches have been able to take advantage of intrinsic-material physical and chemical property

differences, respectively; such would include density, magnetic and electrical conductivity and chemical reactivity.

- PCB scrap is markedly heterogeneous in nature and the key to all mechanical treatment methodologies is in the liberation of the component material fractions. This is somewhat less of a constraint with hydrometallurgical treatment approaches.

5 Sector-based Eco-design

The ultimate goals of the WEEE Directive in respect of the elimination of landfill consignment are essentially to be realised by a combination of recycling, re-use and remanufacturing. To cultivate a hierarchy of recycling and re-use it is apparent that design for disassembly is a key element, in that disassembly has clear impact on overall financial viability and additionally can place limitations on the recycled material yields attainable. Disassembly of WEEE is necessitated for a number of reasons – to remove materials and components which may be viably utilised for re-use or remanufacture, to remove materials having negative environmental impact on subsequent recycling, as exemplified by those within the RoHS Directive (see Section 5.1 below), and to segregate effectively material streams to enhance yields from subsequent recycling-process stages.

There are a number of restrictions on the ultimate recyclability of materials contained within WEEE. As an example, the presence of plastics containing brominated flame retardants (as within RoHS) effectively restricts their re-use capability. It is, however, apparent, as has been noted in the preceding paragraph, that disassembly can in many cases provide effective segregation of these kinds of materials which shredding cannot, and would have to embrace and integrate separation technologies such as electrostatics which would ultimately generate a reduced yield.

In consideration of purely recycling, which cannot be taken totally in isolation from other interfaces of eco-design, as will be discussed later, the hierarchy of preference may be stated as:

- Maintain product
- Recycle sub-assemblies
- Recycle components
- Recycle materials

Further, design for disposal and recycling may also be considered as one of three major elements embraced by environmental design, the others being related to manufacturing and packaging.

Within the pure context of design for disposal and recycling the following may be set down as examples of considerations to be observed:

- The re-use and refurbishment of components and assemblies
- The selection of materials to enable re-use and minimise toxicity
- The avoidance of fillers in plastics

- The identification of materials deployed within components and goods to facilitate re-use and recycling
- The minimisation of the quantity, the number and the colours of the materials used to facilitate recycling
- Attention to design for Materials Separation
- Design for disassembly
- The avoidance of the use of adhesives where possible
- The limitation of contaminant species where possible which may inhibit or impact upon the effectiveness of recycling efficiencies. These would include additives in plastics and coatings such as electroplated finishes
- Maximisation of the use of recycled with virgin material in plastics recycling
- Designing for serviceability

It is considered that design for disassembly may be construed as the key element in the cultivation of a recycling and re-use hierarchy and may be taken as supporting design for recycling.

5.1   Disassembly

In essence disassembly of WEEE is necessitated by:

- The need to remove materials and components for re-use or remanufacture
- The need to remove materials having negative environmental impact. This is instanced by the materials covered within the scope of the RoHS (Restriction of Hazardous Substances) Directive. The RoHS Directive bans the placing on the EU market of new electrical and electronic equipment containing more than agreed levels of lead, cadmium, mercury, hexavalent chromium, polybrominated biphenyl (PBB) and poly-brominated diphenyl ether (PBDE) flame retardants.
- To effectively segregate material streams to enhance yield in subsequent recycling processes.
- The scrap value is higher when dismantling and separation can result in the recovery of more pure-material fractions. It is a commercial pre-requisite that the labour costs for dismantling and separation must be lower than the gained increase in scrap value.

Disassembly, sometimes referred to as demanufacturing or inverse manufacturing, invariably has a high if not total manual element and there are thus finite restrictions imposed on enhancement due to inherent difficulties in introducing automation in a cost-effective manner. Active disassembly techniques in which products are designed to disassemble under an external stimulus will no doubt form an ever-increasing element in disassembly methodology and many forms of active disassembly are currently being researched and developed. To actually design for disassembly, however, whilst being able to enhance automation and active disassembly, has key significance in that it enables many product enhancements to be realised. DfD (Design for Disassembly) would be inherent from product conception and would involve the selection of components, materials and fasteners to minimise the number of components, to standardise on material types and specifications, and to both minimise and simplify fastener types.

In broad terms there have been identified a number of principles within DfD that will facilitate disassembly and these include:

- Use of biodegradable materials where possible
- Providing accessibility to parts and fasteners to support disassembly
- Weight minimisation of individual components
- Use of standardised joints to minimise number of tools for disassembling
- Modular design for ease of parts replacement
- Use of connectors instead of hard wiring
- Use of thermoplastic as opposed to thermoset adhesives
- Use snap-fit techniques to facilitate disassembly
- Design product with weak spots to aid disassembly

It is considered that the choice of fasteners and associated elements is at the crux of design for disassembly whilst the choice of materials may be considered to be at the core of design for recycling. Design for disassembly has the capability to reduce or remove restrictions currently inherent in recycling technologies and thus have significant impact upon both the technical- and cost-effectiveness of WEEE recycling.

5.2 Fasteners

There are many available choices of fasteners, which can enhance disassembly and, in many instances, manufacturing efficiency.

Snap-fit Fasteners

Typically, snap-fit fasteners are easily joined together without the use of tools. There are many types of snap-fit fasteners, ranging from cantilevers and annular snaps to traps and darts. With slight modifications, these connectors can be designed for repeated assembly and disassembly. For example, the inclusion of a flat head top on an annular snap will allow it to be engaged and disengaged repeatedly without the use of a tool. A cantilever hook that is tapered is more likely to withstand repeated assembly and disassembly. Snap-fit flaps found in radio and calculator battery compartments are good examples of fasteners that can be opened and closed repeatedly. The use of re-usable snap-fit fasteners allows a product to be opened easily for repair and upgrades or for disassembly to recover recyclable components when a product is no longer useful.

Moulded-in Hinges

Hinges are useful in providing ease of access to parts of a product for repair or upgrades. There are a number of types of integral or 'moulded-in' hinges. If hinges need to be attached, they should be attached with ultrasonic energy or with plastic rivets made of the same type of plastic to facilitate recycling. Hinges can also be very reliable since some can be flexed up to a million times without failure.

Welding and Energy Bonding

Use of adhesives on plastic parts contaminates the materials for recycling. Welding plastic parts together is a good alternative method of bonding where an immobile strong connection is needed. If the two plastic parts are made of the same thermoplastic material, ultrasonic welding can melt plastic together to form a strong bond. It is important to make sure that plastic parts are of the same resin type before they are welded together. If the parts are made of dis-similar plastic resins, they will most likely not be able to be recycled at the end of product life.

Focused infrared welding is useful in joining thin-walled parts to thick-walled parts without plastic distortion. Solvent bonding, where an organic solvent is applied to the plastic parts, should be avoided since the solvent can act as a contaminant and workers are exposed to a hazardous material.

The impact of fastener choice and selection has given increasing impetus to the design of a number of innovative systems.

The clever 'push-button' fastener on a Dell personal computer cover is one example of a unique and functional product feature. Simply by pressing in on two buttons, one on each side of the cover, then lifting the cover up, the entire cover can be removed without the use of any tools. IBM Corporation uses a dart connector to hold acoustic foam in place on computer front panels instead of using an adhesive. This eliminates exposure of workers to adhesive fumes and the need to dispose of leftover glue as hazardous material. The foam can also be easily detached from the plastic panel for recycling. Some innovative connectors are designed for multiple purposes. In one of the Dell personal computer models, a lever not only acts as a connector for a rack that holds a number of circuit boards, but it also serves as a handle for pulling the rack out for repair, upgrades or disassembly. When the lever is down, it locks to hold the rack securely inside the computer chassis. When the lever is lifted, the lock releases and it can be used as a handle to lift the rack out of the chassis.

A novel approach to addressing the issues relating to disassembly with fas-teners is in an instance where a component or components have high intrinsic value yet it remains economically unjustifiable to disassemble due to the number of fasteners involved. In such a case, the desirable or high-impact components need to be located within an enclosure in order that they may be retrieved by the removal of a minimal number of fasteners. This introduces the concept of embodied disassembly, where components are spatially arranged within an enclosure such that the relative motions may be constrained by the

Table 4 General guidelines for fastener use.

Recommendation	Advantage
Use the least number of different types of connections as possible	Minimises the number of tools that are needed when disassembling for repair, upgrades or disassembly
Use plastic fasteners made from the same resin type as the part	Facilitates recycling
Use fasteners that can be removed without tools	Facilitates ease of repair, upgrade implementation and disassembly for recycling
If metal fasteners are used, they should be of the same head type, be magnetic and have integral washers	Magnetic fasteners of the same head type with washers are easily disconnected, then separated out magnetically during recycling
If screws are used, use coarse heads *vs.* fine heads	Screws with coarse threads take less time and energy to remove

use of lugs or locators integrated into the components, thus requiring a minimum level of fastening. This embodied disassembly approach allows a complete re-use/recycle scenario for all components and the enclosure.

General guidelines for fastener use may be expressed as in Table 4, whilst in the subsequent table is set down a review of common fastener types and their potential impacts, including recycling, based on results from a German test house.

5.3 RFIDs (Radio Frequency Identification Tags)

A key element in recycling of WEEE is in the identification and referencing of product data to enable material and producer information to be accessed. The requirements of the WEEE Directive include:

- The producer must be uniquely identifiable by a mark on the appliance (in exceptional cases, where this is not possible because of the size or the function of the product, the symbol shall be printed on the packaging)
- A separate mark specifying that the appliance was put on the market after 13 August 2005
- A mark indicating necessity for separate collection (European Standard EN50419 January 2005)
- Information for consumers on separate collection and the importance of recycling with regard to WEEE management
- Information for treatment facilities in order to identify equipment and materials
- Producers must achieve the targets set for the rate of recovery and the proportion of re-usable or recyclable components and materials contained in the appliance

Whilst these demands for data collection may be addressed *via* bar-coding, it is far preferable to employ RFIDs, although more expensive, for this purpose as they offer a number of distinct operational advantages. These would include non-line-of-sight reading capability, their ability to be linked with other communication systems and the uniqueness to a specific unit of a transmitted RFID number. It is anticipated that their use during the entire lifecycle of a product will increase substantially in the near future. In use, RFID tags will be used for storing product information as mandated by the WEEE Directive which will allow for more automated and efficient management of information associated with product recycling. It is also anticipated that product marking and RFID technology will provide producers and waste treatment facilities with new service, applications and business opportunities. RFID technology will thus affect the management of product information during the entire lifecycle of a product.

One of the producer's roles will be to maintain and update their own product databases, and information collected on a product may be retrieved using an RFID routing server from the producer's information system in which all relevant information has been stored. Waste treatment facilities and servicing companies can also provide producers with current precise information on matters such as recycling costs and the numbers and types of equipment received for recycling. In addition, RFID technology will afford producers a better opportunity to monitor the recycling of products if resellers identify products at recollection points.

5.4   Active Disassembly

Active Disassembly (AD) involves the disassembly of components using an all-encompassing stimulus, rather than a fastener-specific tool or machine. When designing for active disassembly, we tend to consider the use of smart materials which undergo self-disassembly when exposed to specific temperatures. Shape Memory Polymers (SMPs) and Shape Memory Alloys (SMAs) form the majority of the smart materials used. Often in the form of screws, bolts and rivets, AD fasteners change their form to a pre-set shape when exposed to a specific trigger temperature, which can range from approximately 65 to 120 °C, depending on the material. Taking the example of the screw, the thread disappears when exposed to the trigger temperature, allowing it to fall naturally out of the cavity without any extra stimuli. In some cases an AD sheath is used around a traditional screw where structural integrity or costs are a significant issue.

Designing for Active Disassembly takes into account both the product architecture and fastener selection. It is important to consider how heat will be applied to the fastener (*i.e.* radiation, convection, conduction), and collection of the fasteners when they have been removed from the assembly. If it is not possible to locate the fasteners externally then it may be worth considering a conductive element which allows heat to be transferred directly to the fastener.

When considering fastener collection it is ideal to make the axes of fastener insertion coplanar. Physical component separation using SMAs will require some thought regarding tolerances and the forces that are required to separate joined components. As with any product incorporating AD materials, a mock-up or prototype will allow the manufacturer to determine the optimum level of separation, trigger temperatures, size and number of fasteners, as well as the method of heat application.

5.5 Design Methodology and Resource Efficiency

Within the context considered herein, the choice of materials is clearly a crucial factor in design for recycling, but it should be noted that materials selection has a broader eco-design impact than merely facilitating or impacting ultimate recyclability. This would embrace the designing of products and manufacturing processes for overall maximum resource use efficiency, *i.e.* using the minimum quantity of materials and other resource inputs, such as water or energy, throughout the product lifecycle. Advanced and innovative design approaches involve the substitution of mass-produced manufactured goods with environmentally enhanced product service systems within the context of what may be termed dematerialisation. This ultimately relates directly to what may be seen as an overall waste hierarchy wherein the optimum waste minimisation strategy is avoidance.

Products should incorporate an overall design methodology such that they may be managed in closed cycles to eliminate waste; this may be within natural closed cycles with an ultimate loop-closure, as exampled by composting, or an industrial cycle with closure exampled by recycling. A key principle in the integration of eco-design is to ensure that the consumption of resources (materials, water, energy, *etc.*) is minimised. Resource efficiency is, however, not concerned wholly with materials specification and production processes, but in the consideration of the complete product lifecycle. This, in turn, raises an additional key factor in the impact of the consumer upon product-use and the necessity to design interfaces that encourage sustainable use.

Ultimately, as shown within the waste hierarchy, the disposal of a product at end of life represents an inefficient use of materials and a resource loss. The reduction of materials usage in the first instance, and the recovery of materials for re-use and recycling, will have both financial and environmental benefits. The elimination of materials at source, as in any waste management strategy, will result in reduced impacts throughout the entire product lifecycle.

5.6 Recycling

For the recycling industry, the key role of disassembly is illustrated in the following flowcharts, where the higher-tiered waste hierarchical approach is demonstrably more realisable. Figure 4 depicts disassembly within the context

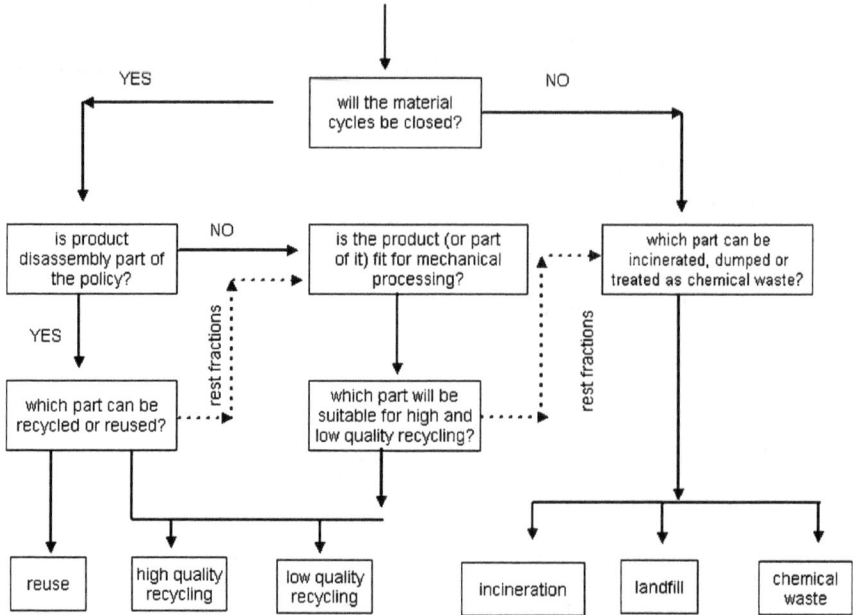

Figure 4   End-of-life destination flowchart illustrating key role of disassembly.

of an end-of-life destination as a flowchart whilst Figure 5 illustrates end-of-life treatment options.

A typical end-of-life route for mobile phones and their ancillaries is shown in Figure 6 to demonstrate both the relative complexity of recyclability deployed and the integration of disassembly within such.

5.7   Constraints on Materials Selection

Note has been made of the crucial role of materials selection in design for recycling. Within established recycling processes, it is important to consider the impact of potentially mixed material feedstock, both in terms of being detrimental to recycling or reducing the value of the recovered fraction with an impurity present. A preferred solution is clearly to design a product without an undesirable either metal or plastic fraction, or alternatively to design the product in such a way as to facilitate separation of incompatible fractions. An example is that of coils and transformers containing copper and iron. These metals are both readily recycled, but the best value is realised when they are in a pure state. Thus for the larger types of these components, they should be made, whenever possible and without sacrificing their electrical efficiency, separable in their pure material fractions. Impurities having impact upon the recycling of metals are depicted in Table 5.

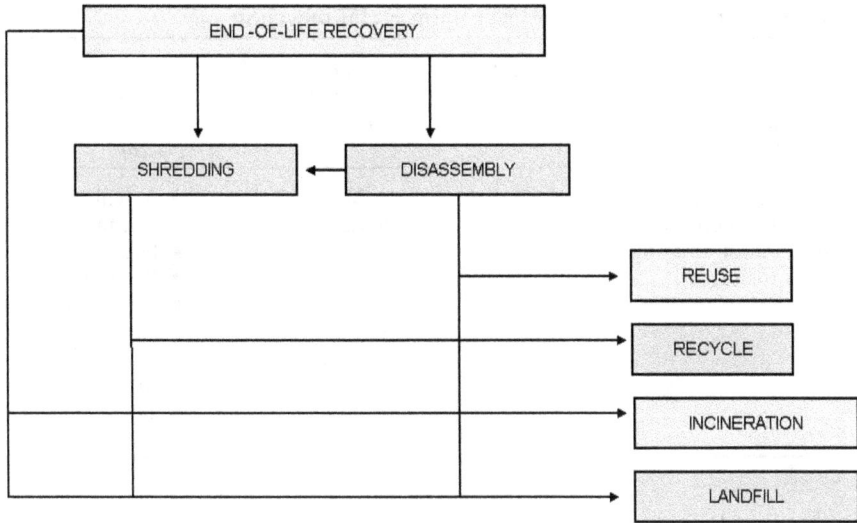

Figure 5 End-of-life treatment options.

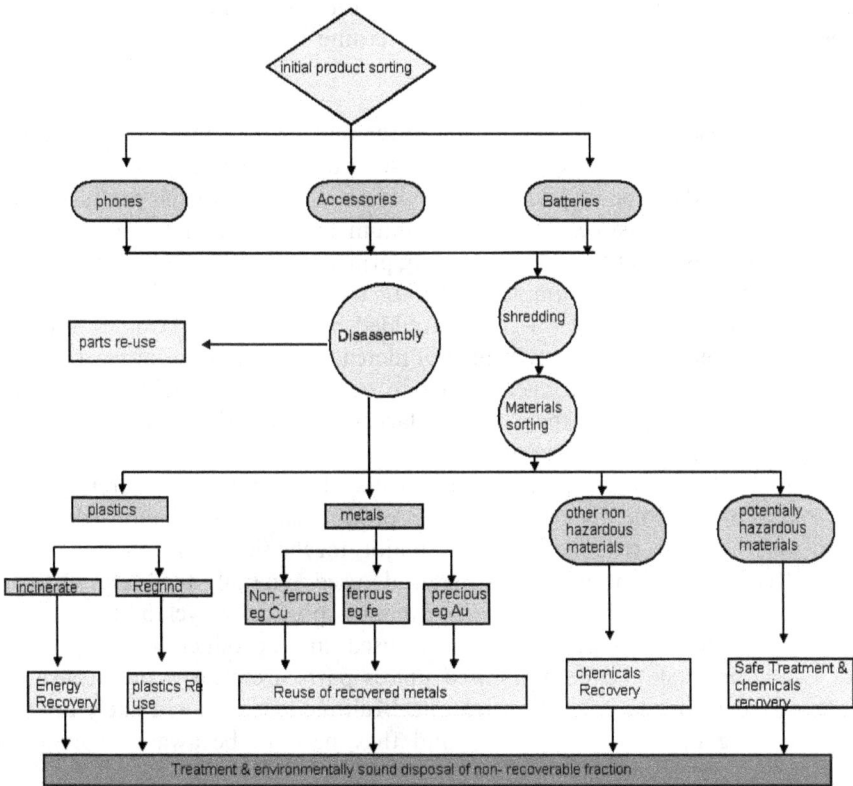

Figure 6 Typical end-of-life route for mobile phones.

Table 5   Impurities having impact upon metal recycling.

For metals the occurrence of certain impurities can be detrimental in their recycling.	*Metal*	*Elements which are detrimental to recycling*	*Elements which reduce the recycling value of the scrap*
In other cases the problem is that only a certain concentration of an impurity can be tolerated	• Copper	• Mercury • Beryllium	• Arsenic • Antimony • Nickel • Bismuth • Aluminium
	• Aluminium	• Copper • Iron	• Silicon
	• Iron		• Copper • Tin • Zinc

5.8   Eco-design Guidelines for Manufacturing

In the selection of materials, in particular plastics, the following design guidelines for both manufacture and end-of-life re-use/recyclability should be observed. General recommendations concerning materials selection are summarised in Table 6.

1. **Minimise material usage.** Using fewer materials to make new products reduces both the use of natural resources and the amount of material that needs to be recycled or disposed of at the end of the product's life. Whenever possible, only the minimum amount of material should be used. Designers can use engineering principles to utilise reduced amounts of material. For example, stiffening ribs, a double wall with tack-off ribs or gas-assisted injection moulded-box beam ribs can be used to increase plastic stiffness instead of increasing the amount of plastic used. Upgrading to a stronger plastic to achieve stiffness will also usually require less plastic than using a larger amount of a weaker plastic to achieve the required strength.

   While meeting product requirements, the number of different plastic and non-plastic materials used in a product should be minimised. Using only one or two materials for major mechanical parts is preferable. Reduction in the variety of materials used generally facilitates efficient disassembly of the product and enhances product recyclability.

   Frequent changes in materials used in a product, or in product upgrades, should be minimised unless parts are clearly and accurately marked for material identification. Multiple recyclers will likely be processing the product material and they need to be aware of material selection changes.

   Changing materials without accurately marking the changed parts could result in contamination of recycled resin.

Table 6 General recommendations concerning materials selection.

Recommendations	*Reasons for Recommendation*
• Use as few different types of materials as possible	• Facilitates sorting of materials for recycling
• Avoid use of dangerous and hazardous substances	• Legal compliance • Reduce risk of contact with hazardous substances during manufacture, use and disposal • Reduced disposal costs
• Avoid using materials characterised as scarce resources	• Limits use of scarce resources
• Use materials which can be recycled within established recycling systems	• Reduces consumption of resources and results in increased value on disposal
• Reduce consumption of materials: over-dimensioning	• Reduces consumption of resources and results in increased value on disposal
• Reduce packaging • Label materials • Ensure that different materials can be separated • Compare packaging alternatives using LCA • Reduce spillage and waste	• Reduces consumption of resources • Stimulates recycling • Reduce consumption of resources

2. **Resin compatibility.** If more than one type of plastic is to be used within a product, recyclability will be enhanced if the plastics are compatible for recycling together. If incompatible resins are to be used, designers should ensure that the materials can be physically separated. It may also be desirable to select resins that have different specific gravities; many recyclers separate materials by differences in specific gravity. To avoid unnecessarily using incompatible resins, material selection options should be discussed with suppliers. Suppliers should also be consulted about recyclability when the use of composite materials is planned. Additionally, suppliers can help ensure use of materials which meet OEM and regulatory specifications.

3. **Recyclable materials.** Designers can facilitate recycling by selecting materials that can be used in internal 'closed loop' recycling processes. Plastic parts and enclosures should be designed to be recycled into the same part or into a different part within the same product whenever feasible. This approach helps provide an outlet for the plastic at the end of its life. Preferences can also be shown for materials that are readily recyclable externally. Because recycling markets and technology are changing rapidly, it is advisable for designers to periodically investigate recycling opportunities.

4. **Contaminants.** Whenever possible, designers should select resins and design techniques to avoid using materials that may become contaminants in the plastic recycling process. Such materials include: labels, adhesives, coatings, finishes and metal fasteners. Examples of preferable design strategies are contained in the sections that follow.

5. **Hazardous and toxic additives.** Engineering plastics contain vital additives such as colourants, fire retardants, stabilisers, plasticisers, reinforcement materials and fillers. The use of additives formulated with hazardous, toxic or banned materials should be avoided, and due consideration should, in any event, be afforded to recyclability problems posed by the use of all additive materials.

6. **Materials cost.** Resins selected should satisfy both functional and processability requirements. Additionally the total manufacturing cost associated with resin use should be considered over the resin cost alone. For example, plastics may offer significant manufacturing, assembly and disassembly cost reductions over other materials because of their multifunction capability. Plastics capability for part consolidation decreases material costs and may reduce use of natural resources. Total lifecycle costs also include the cost of future product disposition. Designers should consider the potential costs associated with managing their product at end of life.

7. **Recycling plastics.** Using recycled material creates outlets for recycled material, providing opportunities for recycling different products and making recycling more economically feasible. The use of recycled material also may conserve the use of natural resources and extends the life of the plastic. When selecting materials, companies should consider using recycled plastic. To facilitate stable markets for recycled materials, as high a percentage of recycled resin as possible should be selected. Requirements of regulatory and certification agencies should also be considered. Companies can also 'close the recycling loop' by re-using plastic recycled from their own products and by not unnecessarily restricting the use of recycled material. Recycled resins may not be considered for appearance parts because of colour-matching issues. When using recycled plastic, companies should ensure that the material meets safety and performance criteria. Resin suppliers can usually provide this information, but in some cases, a material sample may need to be sent to an analytical lab for verification. Some recycled materials are available that have undergone considerable testing.

8. **Basic design concepts.** Properly designed parts can not only enhance the performance of products but also positively impact the disassembly and recycling of the finished product. Rather than over-designing products and parts, designs should meet optimum needs. Simplified designs are not only less costly but also easier to disassemble and recycle. To optimise designs, finite element analysis or computer-aided analysis of moulding flow and cooling criteria can be used. Designers should also define realistic requirements for stiffness and strength. Designers can

consult with resin suppliers, moulders, mould makers and other suppliers for recommendations and input on optimising designs.

9. **Extending product life.** Extending the life of a product delays its replacement and conserves natural resources. To accomplish this, products and parts can be designed to last longer or to be re-used after appropriate servicing. For example, equipment housings can be designed to be cleaned and/or repaired and then re-used for the next model upgrade (Design for Re-use). The life of plastic enclosures, particularly portable product enclosures, can be extended by selecting a plastic material that can best accommodate repetitive reprocessing. Products, parts and components can also be designed to be upgraded as technology changes, extending the technological life of the product and preventing premature discard of the product material. To facilitate removal, upgradeable parts should be designed as sub-assemblies. Additionally, the life of the plastic resin used in a part can be extended through recycling.

10. **Disassembly considerations.** Whilst design for disassembly has previously been detailed as an important element in design for recycling, and indeed within eco-design as a whole, the following comments are specifically attributable to plastic components. Part designs that make disassembly of products cumbersome result in increased labour costs at the end of the product's life. Increased costs decrease the economic feasibility of recycling or re-using product material. Designers can facilitate recycling and re-use of products and parts by choosing assembly procedures that result in efficient product disassembly and separation of differing materials. However, trade-offs between designing for disassembly and designing for maintenance and servicing should be evaluated when applicable. Minimising the variability in type of fasteners used in a product (*e.g.* screws, clips, nuts) speeds disassembly and reduces the number of different tools required. When moulded into a product, breakaway joints and panels can also speed disassembly. The ease of disassembly of a product can be enhanced by lowering the number of separate parts required in a design. For example, multiple parts can be designed into one part. This will reduce the number of fasteners used and thus reduce the amount of time required for disassembly, sorting and recycling. Disassembly may even be eliminated by joining parts made of the same material through alternative joining methods that do not require the use of hinges, fasteners, inserts and other attachment devices.

11. **Wall thickness and strength of parts.** When designing part wall thicknesses, it should be borne in mind that thinning the walls conserves natural resources but may result in trade-offs of strength, stiffness, toughness, warpage and ease of mould filling. Optimising wall thicknesses according to cost/function trade-offs is preferred over thinning walls. Additionally, design engineers should ensure that national and international safety requirements and standards are met. If a product is not designed with a uniform wall thickness, the thick sections should be cored out. When plastic parts are designed with thin walls, part stiffness

can be reinforced by using one of several environmentally preferred design features. Increasing stiffness using such features allows conservation of plastic material. Design features include:

- narrow ribs used to stiffen a flat surface area; a larger number of narrow ribs is preferable to a smaller number of large and heavy ribs
- bosses (protruding studs or pads used to reinforce holes or for mounting an assembly) and/or
- gussets (supporting members used to provide added strength to features such as bosses or walls)

12. **Fasteners and joining.** Whilst the crucial role of fasteners within a Design for Recycling hierarchy has been emphasised, the following points relate in more detail to the fastening of plastic parts and components. The selection of fastening and joining methods can significantly affect the cost of recycling a product and the recyclability of product material. Careful and consistent choices can decrease the time, cost and amount of scrap associated with product disassembly. Generally, fasteners and joined parts should be accessible and easily removed and sorted. The re-use of a resin depends primarily on the purity of the resin or resin blend in the application and on the availability of a market for the recovered material. The cost-effectiveness of plastics recycling decreases when different plastics or plastic and non-plastic materials are joined together. To facilitate product recycling, designers should avoid the intimate attachment of plastic and non-plastic parts, as well as the attachment of parts made from different plastic materials.

Designers should also ensure that fasteners between two different types of plastic or between a plastic and non-plastic part can be accessed for removal during product disassembly. For example, avoid using insert moulding or recessed attachments (*e.g.* pegs, screws, *etc.*) in these applications. Additionally, plastic parts should be joined in a way that leaves no dissimilar material in the recyclable stream (*e.g.* no adhesives or metal from fasteners). One method is to use plastic fasteners made of the same type of plastic as the parts to be fastened, or which are compatible for recycling with the plastic product. Metal fasteners must usually be removed before product parts can be recycled. When liberated from products and parts, ferrous fasteners can usually be magnetically separated from plastic and other product materials. However, many fasteners are not designed for easy removal, so additional effort is required to separate them from other product materials. Naturally, this increases recycling costs. Therefore, when metal fasteners are to be used in a product, carbon or magnetic stainless steel should be preferred over non-magnetic stainless steel, aluminium or brass. When metal hinges are used in plastic products, break points can be provided on parts for easy hinge removal with limited loss of plastic material. Because of the difficulty in their removal prior to recycling, metal rivets should be avoided in product designs.

If metal inserts are to be used in a product, thread-forming inserts are preferable over moulded-in, heat- or ultrasonically inserted metal fittings.

Preferred alternatives to threaded metal inserts include bolts and self-tapping screws. Designers should avoid using irregularly shaped sheet-metal inserts, typically used for electrical components with moulded-in conductors. The use of metal inserts with concentric ring undercuts or deep locking undercuts should also be avoided. Unless installed on breakable bosses, moulded-in metal inserts can be difficult to remove at the end of a product's life. If moulded metal inserts are used, a weak area in the plastic should be included to facilitate break-off of the insert in preparation for product recycling. When using metal drive pins, a through hole should be provided for easy pin removal. A variety of plastic fasteners are available for use. Because of their flexibility, plastic fasteners permit simple, efficient designs. When made of the same type of plastic as the parts they are joining, plastic fasteners may not have to be removed prior to recycling. This elimination of a processing step reduces time, cost and potential contamination of the recyclable material.

Specialised types of plastic fasteners include blind plastic rivets and ratchet fasteners. Blind plastic rivets are available in a number of types of thermoplastics. Plastic ratchet fasteners can be used to join solid or compressible components of varying thicknesses. Plastic ratchet fasteners are composed of a shaft with a smooth head and flexible annular ribs. When pressed into a hole, the fastener's ribs compress and then pop out on the other side. Snap-fits are useful for fastening and joining plastic parts. Snap fits are moulded into parts, reducing the need for separate fasteners and the potential for contamination of recyclable materials from the use of dissimilar materials (*e.g.* metal inserts, adhesives and other fasteners). If designed for disassembly, snap-fits may allow rapid and efficient separation of materials. Tapered snap-fit designs are more likely to withstand repeated disassembly and assembly, facilitating product servicing.

Designers should use moulded-in snap-fits instead of screws or fasteners only where EMI (electromagnetic interference) shielding concerns are minimal (snap-fits may not provide enough pressure on connected parts to ensure adequate conductive continuity in products requiring shielding). Plastic parts can be designed to allow the use of snap-fits if necessary.

Several designs may be used to make integral or moulded-in plastic hinges. Integral strap hinges use a moulded-in strip of plastic to connect plastic parts. Moulded ball-and-socket or ball-grip designs are usually assembled through snap-fits. Knuckle-and-pin designs may use hooks and/or eyes to form the knuckles that serve as the hinge pivot on an axial pin. Certain types of integral hinges have been shown to flex up to one million times without failure. If an integral hinge is to be attached, then bonding, ultrasonic welding or plastic rivets should be utilised. Any plastic rivets used should be of a material similar to the parent parts to facilitate recycling.

Adhesives (*e.g.* glues, epoxies) usually introduce a dissimilar, contaminating material to potentially recyclable parts. This may impact the quality and re-usability of the recycled material for new applications, unless the adhesive can be removed easily or is thermally stable and compatible with the plastic to which it adheres. Designers should either look for adhesives that do not already affect recyclability or consider alternative fastening options.

Several types of welding and bonding methods are available for joining plastic parts. Designers should consider the energy requirements associated with thermal-intensive joining methods. When two parts to be attached are made from the same type of thermoplastic material, ultrasonic welding is an environmentally acceptable method. Ultrasonic welding melts plastics together using ultrasonic energy, forming a strong bond (see Figure 8). When one of the two parts is made from a different type of material, however, the parts will usually have to be separated before being recycled. Separation of materials bonded ultrasonically is very difficult and will most likely render part recycling infeasible.

Focused Infrared Welding creates pressure-tight welds without distorting the plastic being melted. With Focused Infrared Welding, only the mating surfaces are heated and the temperature of the two parts can be controlled, permitting attachment of thin-walled parts to thick-walled parts without plastic distortion.

Solvent bonding is a useful way to bond similar thermoplastic materials. However, the smallest amount of organic solvent needed should be used in order to reduce environmental impact. When solvent-bonded parts are moulded from the same plastic material, they are treated as a single part for recycling because the solvent bond adds no contamination to the recyclable stream.

Swaging and staking may be good substitutes for fasteners used between plastic parts. Environmentally conscious staking involves using thermal or ultrasonic energy to join two parts made of similar thermoplastic material. The parts are joined by reforming the plastic material and forming a locking head. Because no dissimilar materials are introduced to the plastic parts during joining, the plastic is a good candidate for recycling into new applications.

Spring or speed clips provide a rapid and inexpensive method of fastening parts as long as they are easily accessible. Spring clips are usually snapped over a boss, stud or wall section in the plastic part and can be removed by breaking off the boss or stud or by prying with a special tool.

To facilitate recycling of parts fastened with spring clips, the clips should be removed and also recycled when feasible. Metal nuts, bolts, washers and screws, like other metal fasteners, must usually be removed from products prior to recycling. Rapid and efficient removal of these fasteners reduces recycling costs. To reduce separation efforts, integrated washers are recommended. Also, good designs should provide adequate, visible head-accessibility and clearance to allow for a removal tool's range of motion.

Fasteners must be designed to be removed and replaced on products and parts that are likely to be re-used or serviced. Fasteners and joints should also be designed so that they will not be damaged during disassembly and re-assembly. For example, consider designing snap-fits so they can be easily disengaged without breaking.

Because of the possibility of breakage, designers may wish to provide redundant features for fasteners and attachments on products designed to be re-used or serviced. Specific fastening designs and methods may also protect against breakage. Examples include:

- internal hinges that are integral to the plastic part
- use of the same type and size of screw heads within a product and product lines and
- use of metal boss-caps instead of threaded metal inserts (the press-on metal caps fit over plastic bosses and are easy to remove for recycling, plus they provide a guide for starting screw threads back into the hollow boss after product servicing and will withstand repetitive screw assembly and removal without stripping)
13. **Coatings and Finishes.** Unless removed prior to recycling, coatings and finishes can affect the performance of recycled material. The presence of coatings can also affect the appearance and physical properties of products made from recycled material. When recycled plastic is remoulded, small amounts of paint remaining in the plastic can affect the mechanical performance of the recycled material. When plastic resin is recovered from either decorative or conductive-coated parts, the resin may require performance re-evaluation before use in information technology applications.

Integral (moulded-in) finishes are preferred for decorative appearances because they eliminate the need for paint or coatings. Integral finishes are generally cost-effective in addition to being environmentally sound.

Paint and coatings are not only difficult and often costly to remove, but may contribute to the generation of chemical emissions and waste. Residual paint particles can also act as stress concentrators in moulded recycled parts, potentially reducing impact, fatigue and toughness characteristics. Integral colourants are thus preferred over exterior coatings whenever feasible.

Pre-coloured plastics or natural-plus-concentrate are good choices.

If decorative coatings are to be applied, water-based coatings are preferred over solvent-based coatings.

Certain types of decorative finishes are easier to maintain and refurbish on serviceable equipment. For example, low-gloss and textured surfaces hide blemishes and may help eliminate the need for painting. If painting is considered as a refurbishing option, parts should be designed so that the painting will require little or no masking. Masking requires OEMs to use additional resources and often adds cost when the parts are refurbished.

Metallic conductive coatings, including EMI and RFI (radio-frequency interference) shielding, can be difficult to remove from plastic parts. Such coatings may also contribute to chemical waste and air and water emissions, both during use and during the recycling process. If metallic particles are carried over into the recycled plastic, the plastic's physical properties may be affected, reducing its feasibility for use.

Since vacuum deposition introduces only a small amount of metallic material onto a part, it is preferred over metallic coatings such as copper- and nickel-based paint or plating. However, the use of metallic coatings and vacuum deposition should be minimised whenever feasible.

Mechanically attached metal shrouds or metallic foil/plastic (flexible) laminates are preferable over other EMI/RFI shielding methods because they are easier to remove. However, the total cost associated with shielding (including the labour costs for attachment and removal) should be considered in determining shielding requirements.

14. **Material identification.** Once moulded, engineering plastics become very difficult to identify. Testing the material is time-consuming and not always conclusive. Marking can provide critical information to recycling facilities, identifying not only plastic resins but also additives that may necessitate changes in the recycling process. At a minimum, plastic enclosures and significant-sized parts should be marked according to ISO Standard No. 11469. This marking may be enhanced with the addition of another line that indicates the commercial resin name (followed by the word 'resin') or the OEM material code name. Designers may also wish to mark products with additional information that may facilitate product re-use and recycling. Knowing the grade of a plastic material at a part's end of life can facilitate its re-use in equipment and its resale to a recycling vendor.

Products and parts should be marked for identification with the realisation that the OEM may be responsible for managing the materials at the end of the product's life.

A number of methods exist for marking information on plastic parts. Generally, marking through tooling is preferable to marking by labels, pad printing, bar coding or laser inscribing. For information on labels, see the section below. Moulded-in markings are one of the most environmentally conscious marking methods available since they require no use of other materials or chemicals, thus reducing the likelihood of contaminating the recyclable material.

Unless labels and their adhesives are completely removed before a product is recycled, they may introduce dissimilar, contaminating materials into the recycling stream. To reduce contamination from surface labels and adhesives, use moulded-in labels or print labels on the same type of plastic as the part to be labelled and attach them through methods that leave no contamination (*e.g.* ultrasonic welding or solvent bonding). When labels are placed on separate moulded parts, they can also be snap-fitted onto enclosures, allowing easy removal.

15. **Plastic processing.** Encourage plastic processors to select processing methods that are the most efficient, minimise material use and resin scrap, facilitate recycling of the end-product, have the least effect on resin performance and provide the best balance of economics.

Some moulders show more environmental initiative in managing plastic materials than others. OEMs can encourage moulders to use processing methods that produce less plastic scrap and decrease impacts on the plastic's recyclability. For example, OEMs can ask moulders to integrate environmental quality controls with other quality criteria to avoid contamination of recyclable plastic. In the interest of expanding applications for recycled material, OEMs should encourage moulders to increase the usage of regrind back into a product where feasible.

Certain moulding processes facilitate and others may inhibit the recycling of plastic material. Coatings provide finishes to moulded plastics but these finishes can become contaminants during the recycling process.

Alternative processing methods which may decrease the need for coatings and associated pigments include:

- high-pressure thermoforming and counter-pressure structural foam mouldings, both of which can produce parts with acceptable as-moulded surfaces, potentially eliminating the need for coatings and
- in-mould coating processes that limit pigments to the surface layer of a plastic part

Unless made of materials compatible with the plastic to be joined, glues and adhesives typically introduce dissimilar, contaminating materials to a potentially recyclable material. An alternative to gluing two similar materials together is to use a two-shot injection moulding process, particularly when a specific look and 'feel' is desired for the product.

Manufacturers of information technology equipment should also require that moulders use only the type of plastic specified. Changes in material selection can cause significant identification difficulties during product recycling.

Table 7   General recommendations concerning plastics.

Materials, plastics	Reasons for recommendation
• Use as few different types of plastics as possible	• Facilitates sorting materials for recycling • Larger amounts of similar materials increase the value of the scrap

• Choose plastics which:
  ○ can be recycled, *i.e.* thermoplastics (PET, polystyrene) and polyolefins (HDPE, LDPE and PP)
  ○ are compatible on recycling
	• Increases possibility of recycling

• Choose plastics which can be incinerated without emission of hazardous substances • Avoid PVC and other halogen-containing polymers • Do not use brominated flame retardants	• Incineration of plastics is often the most realistic disposal route • In the event of fire in electric and electronic systems, PVC will evolve chlorine and subsequently hydrochloric acid • Legal constraints of RoHS • Toxic substances are emitted during incineration at low temperature. Some of the flame retardants are themselves toxic (PBB, PBDE)
• Avoid adhesive labels on plastic surfaces	• Contaminates material on recycling

When plastics are recycled into new products, the new round of processing adds to the heat history and potential thermal degradation of the plastic material. The following processing recommendations should reduce the heat history and thermal degradation of plastics, facilitating multiple recycling of plastic material:

- use a properly sized moulding machine
- ensure that plastic is not processed above recommended melt temperatures
- reduce plastic's exposure to long heating cycles and
- have moulders consider the use of tooling which minimises regrind

For example, use heated runner systems instead of cold-runner moulds. General recommendations concerning the use of plastics are summarised in Table 7.

References

1. S. Yokoyama and M Iji (Resources and Environment Protection Research Laboratories, NEC Corp., Japan), Recycling of Printed Wiring Board Waste, *Proceedings of the IEEE/Tsukuba International Workshop on Advanced Robotics*, Tsukuba, Japan, Nov. 8–9, 1993.
2. M. Melchiorre and R. Jakob, *Microelectron. J. (IEEE)*, 1996, **28**(8–10), xxii–xxiv.
3. S. Zhang, E. Forssberg and B Bjorkman (Dept of Chemical & Metallurgical Engineering, Lulea University of Technology, Sweden), Metal recycling from electronic scrap by air table separation – theory & application, EPD Congress, *Proceedings of the Session Symposium*, 1998, 497–515, Minerals, Metals & Materials Society, Warrendale, PA. (CODEN: 65TJAT, CAN 128:219715 AN 1998:182587).
4. BIGAT: www.recyclers-info.de/de/bigat/prasengl.htm.
5. K. Saito (National Institute Resources & Environment, Tsukuba, Japan), Recovery of valuable metals from PWB wastes (2) hydrometallurgical treatment of PWB wastes, *Trans. Meter. Res. Soc. Japan*, 1994, **18A** (Eco-materials), 211–214 (CODEN: TMRJE3 Journal written in English, CAN 123-88995 AN 1995:700167).
6. S. Yokoyama and M. Iji (Resources and Environment Protection Research Laboratories, NEC Corp., Japan), *Proceedings of the IEEE/Tsukuba International Workshop on Advanced Robotics*, Tsukuba, Japan, Nov. 8–9, 1993.

European Recycling Platform (ERP): a Pan-European Solution to WEEE Compliance

SCOTT BUTLER

1 Brief Introduction to WEEE

1.1 The WEEE Directive

The production of electrical and electronic equipment (EEE) is one of the fastest growing areas of manufacturing in the world. With broad and expanding consumer take-up of these products, there is also a mounting issue of waste at the end of life.

Across 27 European states, it is estimated that e-waste will rise 2.5 to 2.7% per year – from 10.3 million tonnes generated in 2005 (about one-quarter of the world's total) to roughly 12.3 million tonnes per year by 2020.[1] It is not just an issue of quantity of waste. As some of this electrical and electronic equipment may include hazardous materials (*e.g.* lead, mercury and cadmium), it risks causing environmental problems if the waste is not handled effectively.

In June 2000, the European Commission put forward proposals to address this issue, and in December 2002 these were passed as the EU Waste Electronic and Electrical Equipment (WEEE) Directive.[2]

The major provisions of the WEEE Directive are:

- To make manufacturers (or anyone else putting a product on the market in the EU) liable to pay for take-back, treatment and recycling of end-of-life equipment
- To improve re-use/recycling of WEEE

Issues in Environmental Science and Technology, 27
Electronic Waste Management
Edited by R.E. Hester and R.M. Harrison
© Royal Society of Chemistry 2009
Published by the Royal Society of Chemistry, www.rsc.org

- To ensure the separate collection of WEEE
- To inform the public about their role in dealing with WEEE

EU member states originally had until 13 August 2004 to transpose the legislation into national law. Although few countries actually met the original deadline, WEEE laws are now in place in the major European countries.

Belgium, the Netherlands and Sweden already had in place mandatory take-back of household WEEE predating the EU Directive. In all cases they opted for a monopolistic WEEE take back organisation to take over the obligation of individual producers against payment of various fees. These tended to be set up as not-for-profit organisations by the affected industry sectors.

1.2   Producer Responsibility

Producer responsibility means making the 'producer of a product' responsible for the management of that product once it becomes waste.

1.3   Household and Non-household WEEE

Producer obligations for EEE sold to households are calculated on the basis of sales-based market shares (by weight) in different product categories (see Table 1). For example, if a producer is responsible for 15% of the Large Domestic Appliances sold then they are responsible for 15% of the Large Domestic Appliances that are returned as waste.

For non-household/business products:

- put on the market after 13 August 2005 – producers are responsible for financing the collection, treatment, recovery and disposal of the EEE that they supply to businesses
- put on the market before 13 August 2005 – if WEEE from businesses is being replaced by new equivalent products, the EEE producer is

Table 1   WEEE product categories.

Category	Type
1	Large household appliances
2	Small household appliances
3	IT and telecommunications equipment
4	Consumer equipment
5	Lighting equipment
6	Electrical and electronic tools
7	Toys, leisure and sports equipment
8	Medical devices
9	Monitoring and control instruments
10	Automatic dispensers

responsible for financing the collection, treatment, recovery and disposal when supplying the new products. If WEEE is not being replaced, the end user (the business) has to pay.

1.4 Marking EEE Products

Producers are required to mark EEE with a crossed-out wheeled-bin symbol, a producer identifier mark and a date mark. European standards marking body CENELEC has developed a standard for WEEE Directive marking (see Figure 1).

In some cases, because of the size or the function of the product, the symbol can be printed on the packaging, on the instructions or on the warranty. The wheeled-bin mark aims to help minimise the amount of WEEE disposed of as unsorted municipal (household) waste.

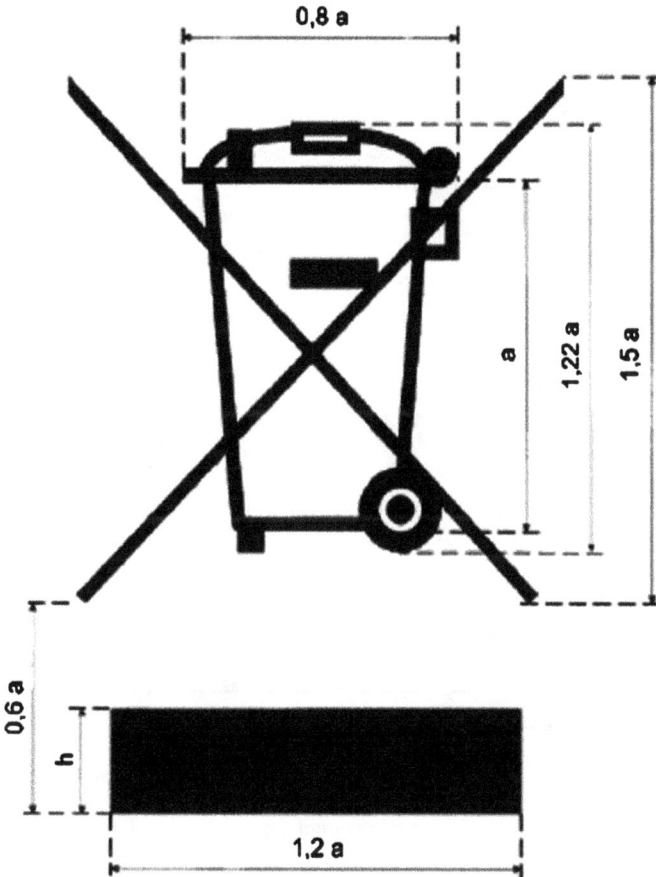

Figure 1 WEEE mark.

1.5 WEEE Collection Points

Although the collection systems (and responsibilities) differ across Member States, the common collection points for WEEE are Civic Amenity sites, Electrical Retailers and own take-back systems operated by EEE producers. These collections points are then served by Producer Compliance Schemes.

1.6 Product Categories and Waste Streams

There are ten product categories that are applied to EEE (see Table 1).

WEEE tends to be collected at Civic Amenity sites in several different streams (see Table 2).

1.7 Producer Compliance Schemes

Producer Compliance Schemes are organisations that co-ordinate the legal compliance of their EEE producer members. Their specific roles and functions differ in each Member State, but they tend to perform the following services:

- Registration of producer members with Enforcement Bodies
- Collection of sales data from members and reporting to Enforcement Bodies
- Developing partnerships with Civic Amenity sites and Retailers to secure access to WEEE
- Managing service partnerships to collect and treat WEEE according to WEEE Directive requirements and member obligations
- Confirming compliance of scheme/members to Enforcement Bodies

1.8 Variations in National WEEE Laws

There are significant variations between how the Member States have interpreted and implemented the WEEE Directive into national legislation. Areas of variation include the requirements for registrations, definitions of WEEE and WEEE weight, the application of product fees to cover the costs of recycling,

Table 2 WEEE collection streams.

Stream	Examples
Large domestic appliances	Washing machines, dishwashers and microwave ovens
Cooling appliances	Refrigeration equipment and domestic air conditioning
Mixed WEEE	Consumer electronics, small household appliances, IT equipment, tools and toys
Display equipment	Televisions and IT monitors
Lighting	Fluorescent tubes and energy saving bulbs

the use of financial guarantees for managing future WEEE and the exact roles of producers, distributors, retailers, government, businesses and householders.

For a detailed analysis of the transposition of the WEEE Directive across Europe see the Perchards Report entitled 'Transposition of the WEEE and ROHS Directives in Other EU Member States'.[3]

2 Introduction to European Recycling Platform (ERP)

2.1 European Recycling Platform

European Recycling Platform (ERP) was formally created in 2002. It was formed as the first ever Pan-European Producer Compliance Scheme. Its primary objective is to help effectively implement the WEEE Directive.

The mission of ERP is 'to ensure cost effective WEEE implementation for the benefit of consumers and our members, and to promote Individual Producer Responsibility'.

2.2 Founder Members

The founder members of ERP are: Braun Gillette, Electrolux, Hewlett-Packard (HP) and Sony.

Braun Gillette is now part of P&G and continues to manufacture a wide variety of products, including electric shavers, oral care products, beauty products and household appliances.

The Electrolux Group is the world's largest producer of powered appliances for kitchen and cleaning use, including refrigerators, washing machines, cookers and vacuum cleaners.

HP is a leading global provider of products, technologies, solutions and services to consumers and businesses. The company's offerings span IT infrastructure, personal computing and access devices, global services and imaging and printing.

Sony manufactures audio, video, communications and information technology products for the global consumer and professional markets.

2.3 Timeline

2000–2002:	Co-operation between founder members in lobbying of EU Directive
Sept. to Dec. 2002:	Development of Concept and Co-operation Agreement
11 Dec. 2002:	Notification to DG Competition
16 Dec. 2002:	Press announcement of ERP

Initial reactions from industry and government to the formation of ERP were positive. They included expressions of interest from more than 20 possible member companies, and the full support from the EU Directorate General Competition and Environment.

2.4 Founding Principles

ERP had a number of founding principles:

Collection and Take-back Requirements. Producers should not be made responsible for collection from individual householders.

Financing Historical Waste. Producers should have the option to communicate their cost to the markets, but legislation should not require mandatory fixed uniform fees, nor that these fees be transferred to a central/single system.

Mandatory Registration. All producers need to be registered and issued with a unique registration number by each National Register before being able to sell a product.

Data. National Registers should collect sales (tonnage) and collected materials (tonnage) data on a regular basis, at worst enabling a quarterly statistical analysis.

Competition in Compliance Systems. All Member States should implement competitive compliance systems that enable producers to have a choice between producer compliance schemes, and promote competitive pressures in all aspects of the system.

Individual Producer Responsibility. Individual Producer Responsibility is a policy tool that provides incentives to producers for taking responsibility of the entire lifecycle of his/her own products, including end of life.

2.5 Structure

ERP developed and operates a common waste management procurement platform designed:

- to meet the specific requirements of electrical and electronic producers
- to promote cost-efficient and innovative recycling strategies, while actively embracing the concept of individual producer responsibility as set out in the EU Directive
- to open up opportunities for pan-European recycling services and cross-border competition in the waste management service market

2.5.1 Original Structure

Representatives from each of the founder members sat on a Steering Committee (see Figure 2) to oversee the actions of:

- National Teams – local, founder-member based teams to manage the formation of ERP against the backdrop of national implementation of the WEEE Regulations
- Sub-committees – founder-member based operations, communications and legal teams

2.5.2 Initial Work Plan

The founder members then developed and delivered a work plan to enable ERP to form, select operational partners and to ultimately become a stand-alone entity. This work plan (see Figure 3) included:

- Estimating annual WEEE volumes to be managed by ERP across Europe – approximately 500 000 tonnes per year
- Identifying potential service partners (management, logistic, recycling) – 15 possible companies were identified and asked for indicative offers
- Calculating realistic cost-saving potentials on the basis of indicative offers compared to existing schemes/operations
- Completing final tender exercise – December 2004

2.5.3 New Structure

With the transition from start-up to ongoing operations complete, ERP's Board approved a new management structure in January 2007 (see Figure 4).

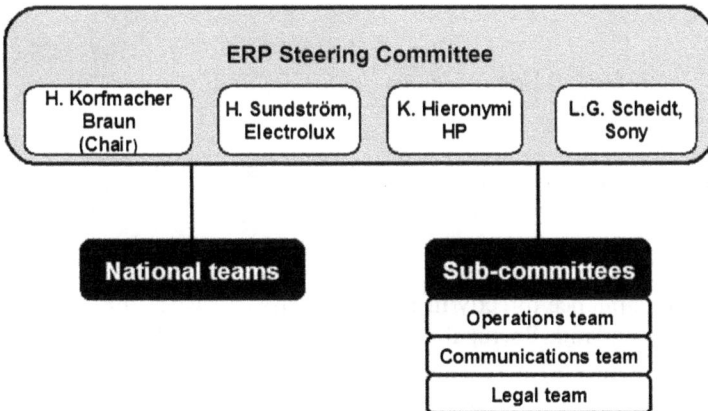

Figure 2 Original ERP structure.

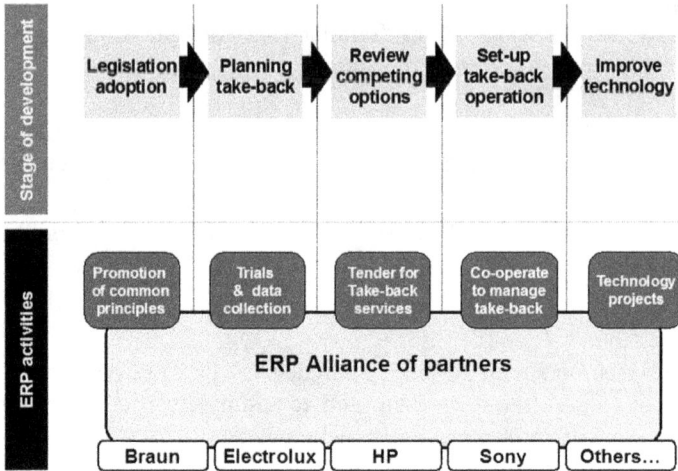

Figure 3 ERP work plan.

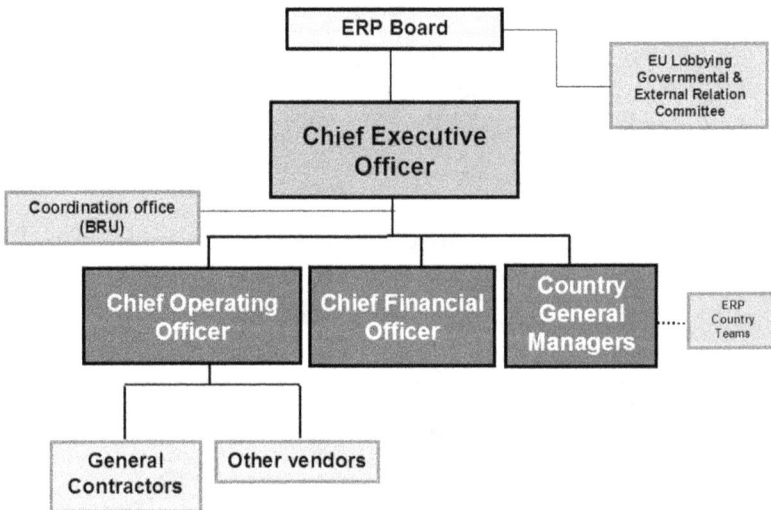

Figure 4 New ERP structure.

2.6   Scope of services

ERP offer compliance directly in nine countries – Austria, France, Germany, Ireland, Italy, Poland, Portugal, Spain and the UK (see Figure 5).

2.7   The Operational Model – General Contractor Approach

ERP is managed with the same business focus as that used by its members to build their own market leadership positions. It is a lean structure that

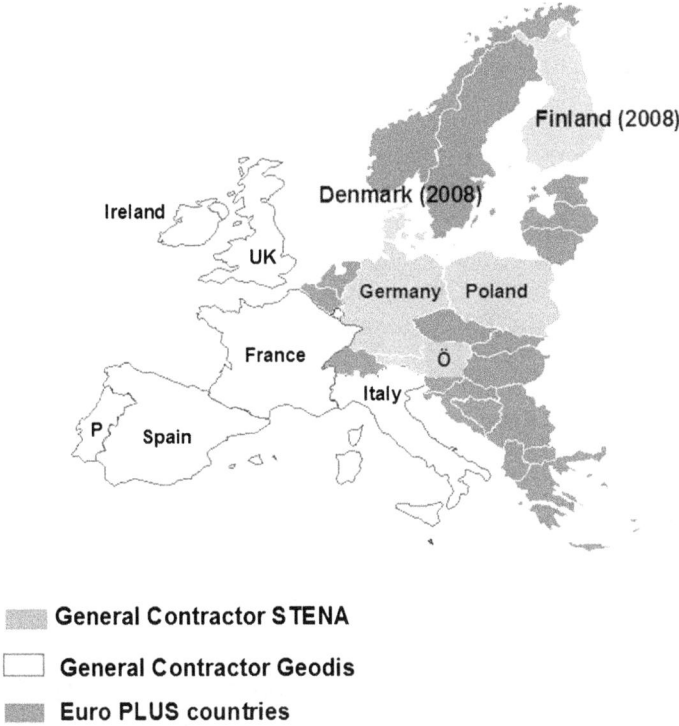

Finland (2008)

Ireland

Denmark (2008)

UK

Germany   Poland

France

Ö

Italy

P   Spain

▓ **General Contractor STENA**

☐ **General Contractor Geodis**

▓ **Euro PLUS countries**

Figure 5   ERP countries.

outsources all its services, which enables:

- Operational flexibility, a prerequisite to constant process improvement
- Sharp reactivity to market opportunities to maintain its competitive position

The General Contractors operate and manage all operations on the take-back process. They will serve any collection point where necessary to reach compliance. Operations, collection, recycling and reporting are currently handled by two General Contractors, both carefully selected after a strict tendering process.

Geodis Valenda is currently the General Contractor for France, Ireland, Italy, Portugal, Spain and the UK. It is a French-based Reverse Logistics specialist.

Stena Metall is currently the General Contractor for Austria, Germany and Poland. It is the Nordic-based provider of a range of recycling and environmental services.

ERP operates according to a set of core principles that are fundamental to the protection of consumers, business and the environment. These are monitored on a regular basis, based on data (volume collected and treated) and Key

Performance Indicators (recovery and recycling) as well as process audits to assess its service performance. The collection and validation of key performance data is a constant process.

2.8 Euro PLUS

ERP has expanded its compliance services through an agreement with 1WEEE Services GmbH. ERP now offers its European members the possibility to reach beyond the current nine countries where it operates directly through its General Contractors and ensure local compliance in an additional 20 countries by subscribing to the Euro PLUS Package.

These services include:

- Advice on WEEE registration
- A connection to national recycling solutions (outside ERP coverage)
- Reporting
- Legal and business consulting
- Financial guarantees

3 ERP in Operation

ERP has direct operations in nine EU Member States (see Table 3).

3.1 Country Summaries

ERP therefore has a significant role in all the EU Members States in which it operates, with a market share ranging from 9% to 30%. The strategy was always to be big enough to be flexible and competitive, without dominating in any one market or Member State.

3.2 Key Performance Indicators

Each ERP operation collates and analyses sets of Key Performance Indicators. The top level of these is the WEEE collected data (see Table 4).

With the implementation of the WEEE Directive phased-in across the EU, the ramp-up of operations has been significant, with volumes doubling in 2007 compared to 2006. With Italy going live in 2008, the UK operating for a full calendar year, and systems settling and expanding in other Member States, this rise in WEEE collected and treated will continue.

3.3 Members

From 464 producer members in 2005, by November 2007 the membership of ERP was 1100, an increase of 228%. In the same period, the number of pan-European members increased by 42%, from 19 to 26.

Table 3 ERP country comparison.

	Austria	France	Germany	Ireland	Italy	Poland	Portugal	Spain	UK
ERP register for members	Yes	Yes	Yes	No	Yes	No	No	Yes	Yes
National registration fees	No	No	Yes	Yes	Yes	Yes	Yes	No	Yes
ERP registration fees	No	No	Yes	No	No	No	No	No	No
ERP report for members	Yes	Yes	Yes	No	Yes	No	No	Yes	Yes
Number of schemes	4	4	<20[a]	2	14	5	2	7[b]	37
ERP Members	23	406	29	77	100	123	405	87	58
ERP market share	21%	10.5%	15%	22%	13%	9%	30%	11.5%	15%

[a]Germany does not operate a compliance scheme based system but has numerous service providres providing compliance-scheme-like services.
[b]Spain has seven schemes but three act jointly under the same management.

Table 4   ERP Key Performance Indicators – WEEE collected 2006 and 2007.

ERP entity	WEEE collected (tonnes)	
	2006	2007
Austria	12146	11983
France	256	31342
Germany	58270	63119
Ireland	10491	9092
Italy	0	0
Poland	386	551
Portugal	515	9089
Spain	796	18674
UK	0	24761
ERP total		
LDA	10508.6	39475.8
Cold	16963.8	31019.2
CRT/displays	37796.4	62554.0
Mixed/IT/SDA	13843.5	35480.3
Lamps	9.2	81.0
	79122	168610

4   ERP – Beyond Compliance

ERP actively takes an interest in issues beyond the basic operations of WEEE systems. Two examples of this are the long-standing commitment to Individual Producer Responsibility, and a recent survey commissioned by ERP on the new WEEE system in the UK.

4.1   Implementation of Individual Producer Responsibility (IPR)

Making each producer responsible for financing the end-of-life costs of their own-branded products enables end-of-life costs to be fed back to the individual producer. By modifications to the product design, the producer can directly influence the end-of-life cost. Without Individual Producer Responsibility these incentives for design improvements are lost.

Article 8.2 of the European WEEE Directive establishes individual producer responsibility for the recycling of products put on the market after 13 August 2005.

Analysis has shown that ten Member States (Bulgaria, Denmark, Finland, France, Greece, Latvia, Portugal, Slovenia, Spain and UK) have omitted the requirements of Article 8.2 in transposing the WEEE Directive into their national law. Instead, the legislation in these countries makes producers jointly responsible for the recycling of future products, making it impossible to implement individual producer responsibility.

The EC Treaty obliges each Member State to implement the WEEE Directive in such a way as to give full effect, in legislation and in practice, to the wording, object and purpose of the WEEE Directive and not to put in place any measure

Table 5 Analysis of the transposition of IPR in the EU.

Member States which have transposed Article 8.2	Belgium, Cyprus, Czech Republic, Estonia, Ireland, Italy, Lithuania, Luxembourg, Malta, Romania, Slovakia, Sweden and the Netherlands
Member States which have inadequately transposed Article 8.2	Bulgaria, Denmark, Finland, France, Greece, Latvia, Portugal, Slovenia, Spain and the UK
Member States which have partially transposed Article 8.2	Austria, Germany, Hungary, Poland

that would jeopardise the attainment of the Directive's objectives. It is therefore crucial that the EU institutions and the Member States ensure that IPR is correctly transposed and implemented in national legislation (see Table 5).

ERP members are currently working with other producers, academics and technical specialists to identify explore and develop practical solutions to IPR.

4.2 ERP UK WEEE Survey

ERP UK conducted a web-based survey between 28 February and 11 April 2008. It was completed by 51 representatives of producer organisations and 36 representatives of local authorities from throughout the UK. The WEEE Directive 'Looking back, looking forward' took a look at the UK WEEE system nine months on from implementation.

4.2.1 Government Support for Producers and Local Authorities
The survey asked respondents how they felt about the quality of information and support they received from the Government since the Directive's implementation in July 2007. Between one-quarter and one-third said they felt the Government's communication and support were poor. But somewhere between two-thirds and three-quarters felt the information and support they received from Government was either adequate or better than adequate (see Figure 6).

4.2.2 Producer Support for Local Authorities
Around two-thirds of Local Authority respondents said the information and support they received from Producers and Compliance Schemes was either excellent or reasonable, and less than one in five said it was poor (see Figure 7).

4.2.3 Public Awareness and Encouragement
The survey asked respondents if the Government should do more to encourage household recycling of electrical goods, and the answer was a resounding 'yes' (see Figure 8).

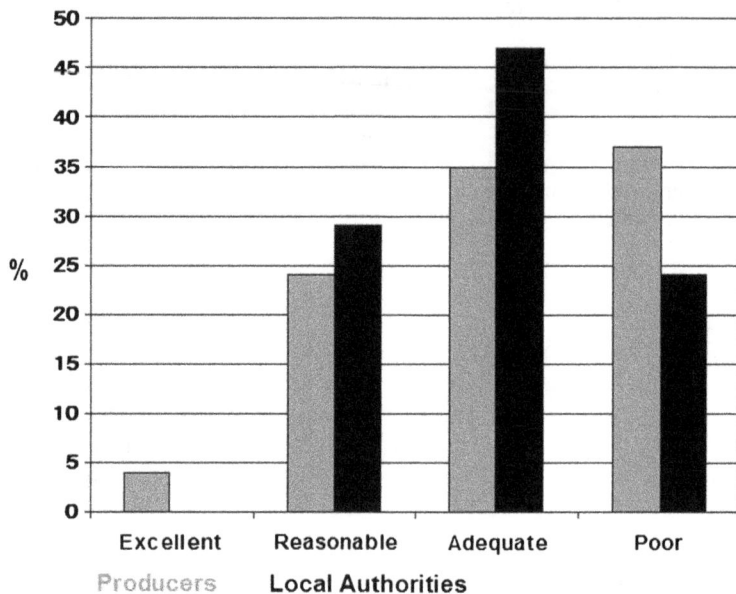

Figure 6 How would you describe the quality of information and support you received from the Government in relation to the WEEE Directive since it was implemented in July 2007?

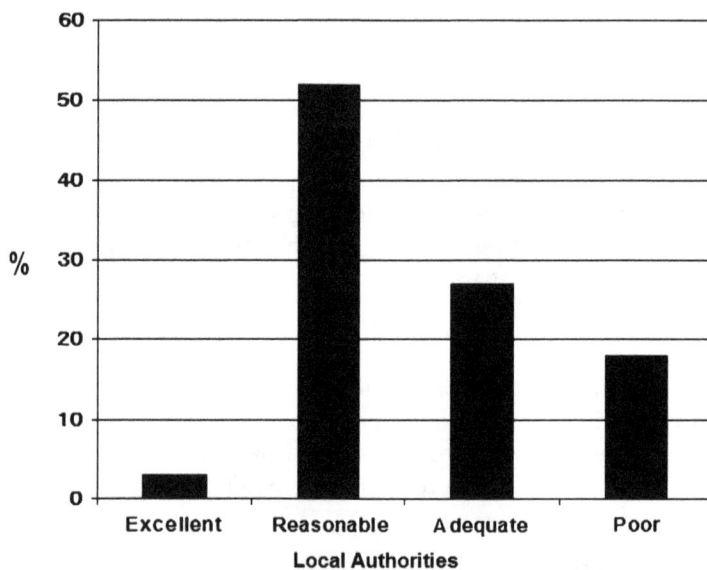

Figure 7 How would you describe the quality of information and support you received from Producers and/or Producer Compliance Schemes in relation to the WEEE Directive since it was implemented in July 2007?

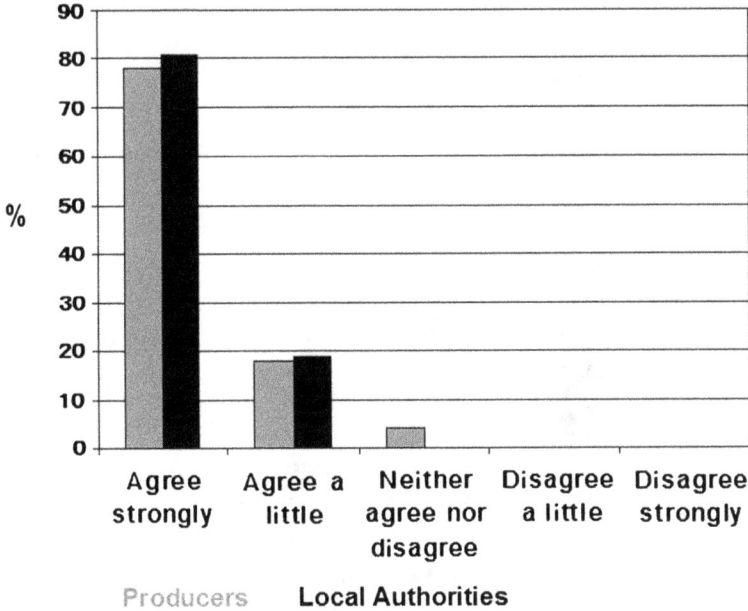

Figure 8 Should the Government do more to encourage householders to recycle unwanted electrical goods?

4.2.4 Individual Producer Responsibility
More than two thirds of both groups agree that a WEEE system in which each producer is responsible for its own products at the end of their useful lives would have a significant environmental benefit (see Figure 9).

5 Summary

ERP's main founding objectives remain:

- to deliver compliance to the ERP members at high quality and best price
- to drive costs down by establishing competition between compliance schemes
- to ensure competitive advantages for its members
- to promote full enforcement of Individual Producer Responsibility

5.1 Key Achievements

- ERP has acted as a catalyst to the development of competitive compliance schemes and most EU Member States now have multiple market players
- ERP has been instrumental in driving prices down, causing a beneficial financial impact on producers, consumers and consequently on national economies

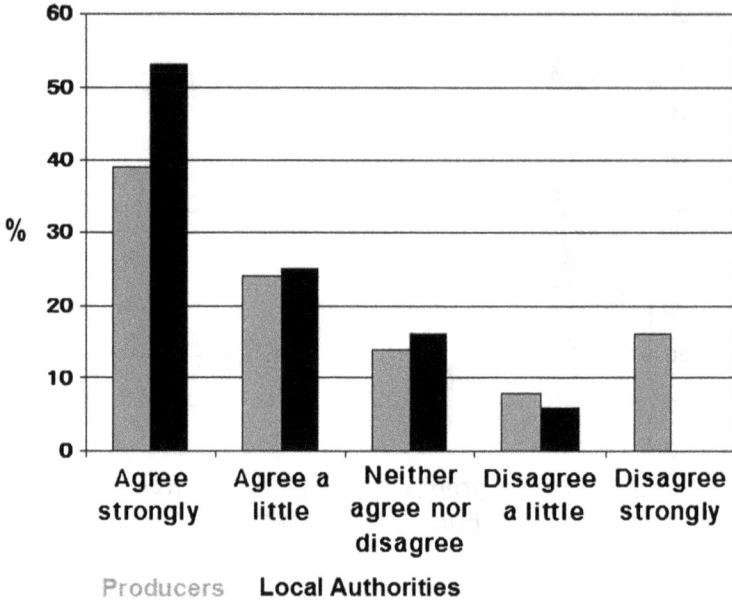

Figure 9 The introduction of a WEEE system in which each producer is responsible for its own products at the end of life would have a significant environmental benefit.

- With an average market share of more than 15%, ERP has reached a size in all countries which allows it to work on a good economy of scale while still ensuring competition
- As of May 2008, ERP has 28 pan-European members, over 1200 members across the national entities, and has processed over 300 000 tonnes of WEEE
- The EU Directorate General Competition sees ERP as a reference in initiating and driving competition in the compliance scheme market and quoted ERP in their guidance document (9-05) about 'competition in the compliance scheme market'
- ERP is already recognised as one of the best-performing compliance schemes in a number of countries

Table 6 shows the importance of competitive compliance systems. Across the product categories, the average competitive system delivers a significantly lower cost than the average monopoly system.

In only five years, ERP has evolved from a good idea to a successful, competitive and fast-growing WEEE management organisation.

The initial objective was to shake the market rules, introduce competition and be the first alternative to the established national recycling management consortia. ERP did this, opened the market and other competitors are now following.

Table 6 Impact of competitive compliance systems.[4]

WEEE fees per product sold	Cooling	Washing machine	Food processor	Notebook	DVD player	CRT monitor/ TV
Average monopoly	16.96€	8.11€	0.93€	2.26€	2.69€	6.30€
Average competition	10.36€	1.66€	0.66€	0.43€	0.56€	3.88€

5.2 Final Thoughts: Interviews with Two Founding Members

In December 2007, ERP celebrated five successful years of operation, and is looking forward to continuing its strong growth trend. In order to gain some perspective on how the last five years have impacted ERP, I spoke with two of the founding members. I wanted their perspective on how ERP has developed, what sets it apart, and where ERP and the WEEE Directive are heading in the next five years.

Klaus Hieronymi, Environmental Management Director at Hewlett-Packard. As one of the original founders of ERP, Klaus Hieronymi, the Environmental Management Director EMEA at HP, has seen at first hand the developments over the last five years. He says, 'It's been amazing to see how ERP evolved from an idea into its existence as a fully operational, serious player. Of course, the way has not always been easy, and ERP has had to tackle numerous, diverse challenges. Yet ERP has emerged as a serious contender, here to stay.'

Hieronymi finds that beyond ERP's pan-European scope, another way in which it is different from other compliance schemes is its embrace of competition on the market. 'ERP has always worked to make the market more competitive, and I believe that's really an aspect which sets ERP apart,' he commented.

ERP was started as a response to the WEEE Directive, which has certainly had an impact on consumers, businesses, and compliance schemes. Hieronymi does feels that the Directive has reduced the amount of electronics and electrical equipment which have entered into the waste stream, but thinks this might have happened even without the EU Directive. 'The prices of raw materials have skyrocketed the past few years, and based on these economic factors, the level of recycling would probably have increased anyway,' he says. Furthermore, the WEEE Directive is due for revision in 2009 or 2010. 'I would like to see that the revised directive considers the aspects of harmonisation. Right now, we have 26 different versions of WEEE take-back legislation, in each member country. Also, it would be nice if the revised legislation provides additional motivation for the producers to recycle, which is lacking in the legislation now.'

But ERP's developments are far from over. In the coming five years, Hieronymi envisages ERP expanding in terms of regions, waste and scope. Opening operations to encompass countries beyond the nine countries where ERP currently operates is one option. Another option is to differentiate the services ERP offers into basic and premium services, for example. Hieronymi also mentions that he can imagine ERP branching into a wider scope, such as battery take-back. 'Finally, it would be great if we could establish a very intense tracking system, which would take the tracking of the waste to a much deeper level than is required by law.'

Hans Korfmacher, Director of Environmental Affairs Europe at the Braun-P&G Company.   Hans Korfmacher can also offer an excellent perspective of how ERP has evolved, as he has been involved since 1995. He says that ERP has evolved from a joint procurement platform into a truly pan-European take-back operator. This is a major organisational step, and it has achieved a high level of competition in the take-back market. 'ERP has helped to create a take-back market, and this model is starting to spread into other take-back markets,' says Korfmacher.

When asked about the aspects which make ERP stand out from other compliance schemes, he lists three main areas: first, ERP is extremely customer oriented. He says: 'Some of our members have come to us from other compliance schemes, and they tell us that they genuinely appreciate ERP's transparency and flexibility. Their needs as a customer are answered.' Secondly, the ERP key-account structure in nine different companies is unique. Under this centralised option, ERP acts as the best solution and offers the most centralised operations. Finally, the third aspect which sets ERP apart is its 'proven track record of being the best performer in some markets. In fact, ERP is actually over-performing in some markets,' according to Korfmacher. ERP has helped to change the authorities' perception that only having one system is the best way to respond to the WEEE Directive. Thanks to ERP, the authorities recognise that having more than one system, and a competitive one at that, offers great advantages to the consumer.

Korfmacher is also enthusiastic about how the WEEE Directive has developed and will develop, echoing the sentiments of Hieronymi. 'The WEEE Directive was intended to manage waste streams from waste electrical appliances, but due to the changes in the global resource market, WEEE has become very valuable. The level of recycling probably would have resulted in a developed take-back market for some categories of WEEE with or without the WEEE Directive,' says Korfmacher. He did mention that one major impact which the WEEE Directive has had is to help the consumer realise that WEEE is something valuable to be recycled and should not end up in landfill. He says that in the coming years the WEEE Directive will continue to change the paradigm of waste products to be handled less as an unwanted burden and more as a resource. 'It all comes down to managing our resources in Europe,' says Korfmacher.

As for how ERP will develop for the next five years, Korfmacher sees ERP continuing to develop into areas such as batteries, packaging and more. 'In general, ERP will continue to develop its pan-European service offering to a pan-European market,' he sums up. To this end, they will need to focus more on the pan-European competitors and less on their national competitors.

References

1. United Nations University, 2008 Review of Directive 2002/96 on Waste Electrical and Electronic Equipment (WEEE).
2. Directive 2002/96/EC of the European Parliament and of the Council of 27th January 2003 on waste electrical and electronic equipment (WEEE) (a) as amended by Directive 2003/108/EC of the European Parliament and of the Council of 8th December 2003 on waste electrical and electronic equipment (WEEE) (b).
3. http://www.berr.gov.uk/files/file29925.pdf, accessed May 2007.
4. R. Veit, November 2005, How do WEEE get it right?, http://www.emsnow.com/npps/story.cfm?id=15184, accessed 20 April 2008.

Liquid Crystal Displays: from Devices to Recycling

AVTAR S. MATHARU AND YANBING WU

1 Introduction

Liquid crystals are special substances that exhibit an intermediate state of matter that exists between its crystal and liquid states. The intermediate state of matter is termed a mesophase and possesses properties of both a crystal and a liquid, and some others unique to the mesophase. The ability of a mesophase to modulate light (either transmit or block) in response to an external applied field is the basic premise of its modern-day application in a liquid crystal display (LCD).

Liquid crystal displays (LCDs) are a technological revolution that impacts our daily lifestyle through interaction with calculators, mobile phones, laptop computers, satellite-navigation systems and the latest flat-screen large-area LCD televisions. Compared with cathode ray tube (CRT) displays, flat panel LCDs are space-saving, lightweight and portable. They do not emit harmful radiation, making them the number one choice in like-for-like applications. For example, LCD computer-monitor sales and shipments equalled and overtook those of CRT monitors in 2004. CRT technology is in decline and revenues have dropped from $10.2 billion in 2004 to a mere $6.6 billion in 2006.[1]

In 2008 the global flat-panel display industry was valued at approximately $115 billion and is forecast to reach $200 billion by 2016 (see Figure 1). In 2008, LCD technology dominated the flat-panel display market with a market share in the region of 90% and sales revenues exceeded $100 billion. The biggest area of current and future growth is LCD TV. In 2006, sales revenue was $22.1 billion and is forecast to increase to $41 billion in 2011.[1]

Issues in Environmental Science and Technology, 27
Electronic Waste Management
Edited by R.E. Hester and R.M. Harrison
© Royal Society of Chemistry 2009
Published by the Royal Society of Chemistry, www.rsc.org

Flat Panel Display Trend

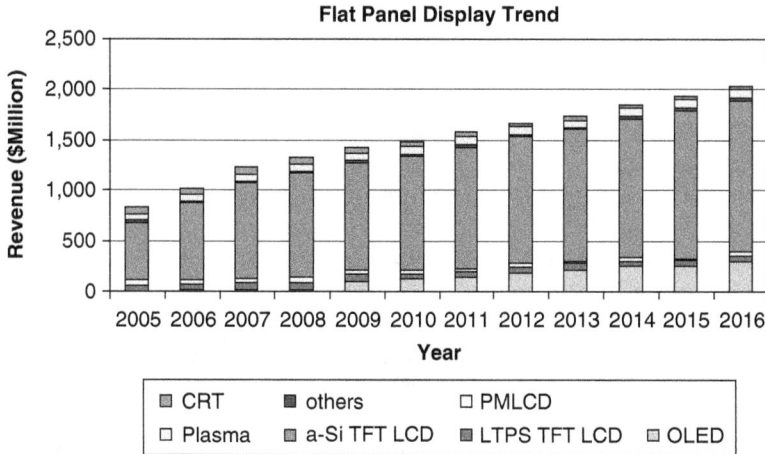

Figure 1 Flat panel display trends 2005–2016 (source: UK Displays and Lighting Network).

There are approximately 1.9 billion TVs in use worldwide, of which 8% (152 million) are LCD. Less than 20% of the TVs in North America and Europe are LCD,[2] with even fewer in the rapidly emerging Chinese and Indian markets. The Chinese consumer is expected to invest in a new generation of digital-ready TV over the next four years. Of the next-generation technologies available, LCDs are going to have the greatest impact, reaching possibly more than 50 million units in 2010 from 16 million units in 2008.

The rapid success of LCD and under-pinning technologies, coupled with a dramatic increase in demand for large area LCD TV now sees the emergence of state-of-the-art Generation 8 fabrication facilities capable of handling 2160 × 2460 mm glass panels, from which eight 46-inch wide panels can be cut. The industry continues to progress and Sharp, in alliance with Sony, recently indicated plans to build the world's first 10th generation (2850 × 3050 mm) facility by 2010, capable of yielding six 65-inch panels, nine 50-inch panels or fifteen 40-inch panels.[2]

As LCDs continue to penetrate and saturate the large-area flat-panel display market there are major global environmental concerns both during their manufacture and now, perhaps more importantly, at their end of life (EOL). Already, LCD-containing waste electrical and electronic equipment (WEEE) has been identified as one of the fastest growing sources of waste in the EU, increasing by 16–28% every five years. A plethora of environmental regulations are now in place, stemming from the inception of the Basel Convention[3] in 1992, namely: WEEE (Waste Electrical and Electronic Equipment);[4] RoHS (Restriction of Hazardous Substances);[5] REACH (Registration, Evaluation, Authorisation and Restriction of Chemicals),[6] as well as local by-laws. The EU WEEE Directive[4] and RoHS Directive[5] are most pertinent to LCDs. The EU WEEE Directive (2002/96/EC) prohibits outright shredding, landfilling or incineration of mercury-containing backlit LCDs with a surface area greater than $100\,cm^2$. LCD-containing WEEE must be isolated and de-manufactured, with targets set for computer-related equipment at 75% for recovery and 65% for re-use.[4]

Detailed studies have shown that the liquid-crystal mixture itself is non-hazardous and the volume per display is minuscule, so that incineration is feasible.[7] However, the expected future volumes and their persistence in the environment may be cause for concern in years to come. Most certainly, the mercury-containing backlights must be removed or treated effectively as mercury is hazardous.

The LCD-manufacturing industry is moving towards greener principles. The main producers of TFT (thin-film transistor)-LCDs are concentrated in China (mainland and Taiwan), Japan and Korea. In 2001 these three countries established the World LCD Industry Cooperation Committee (WLICC) to establish measures to reduce greenhouse gas emissions. The etching and chamber-cleaning processes used during array manufacture are intensive in the use of perfluorocompounds (PFCs) (CF_4, SF_6, NF_3 and CHF_3). Perfluorocompounds have a much higher global-warming potential (GWP) than the equivalent amount of carbon dioxide (CO_2, GWP of 1). The GWPs of SF_6, NF_3, CF_4 and CHF_3 are 22 200, 10 800, 5700 and 12 000, respectively. WLICC works closely with the Japanese Electronics and Information Technology Association (JEITA), the LCD Industries Research Committee (LIREC), the Electronic Display Industrial Research Association Korea (EDIRAK) and the Taiwan TFT-LCD Association (TTLA). The WLICC reached a consensus on reducing the aggregate absolute perfluorocompound emissions to the equivalent of less than 0.82 million metric tonnes of carbon equivalent (MMTCE) by 2010. Many industries have adopted high-temperature catalytic incineration followed by aqueous quench rendering of SF_6 and CF_4 into SO_3 and HF, respectively, which are subsequently removed by alkaline scrubbers. In 2003, WLICC total PFC emissions were 0.899 MMTCE, with contributions from each member organisation as follows: TTLA, 0.266; EDIRAK, 0.377; LIREC/JEITA, 0.256 MMTCE.[8] In 2005, the TTLA's total PFCs in emissions rose to 0.4 MMTCE due to rapid expansion of the LCD TV market.

In 2006 Sharp (Kameyama, Japan) opened the first Generation 8 LCD manufacturing facility designated as a 'super green facility'. The plant comprises its own energy-supply system based on green technologies, such as implementing solar panels on all roofs, aiming to provide one-third of the total electricity required, and all water used in the plant is recycled. Similarly, Sony Corporation recommends its suppliers become a 'Sony Green Partner' in order to meet Sony's own environmental regulation (SS-0529) in addition to getting an ISO14001 mark. In 2003, Merck Ltd, Japan LC Division, acquired 'Sony Green Partner' status to supply liquid-crystal mixtures to Sony Semiconductor Kyushu Co., Ltd, Japan. Partnership status is reviewed bi-annually in order that high environmental standards are maintained throughout.

The purpose of this chapter is to give information on liquid crystals, LCDs, their manufacture and de-manufacture, their market value, their environmental

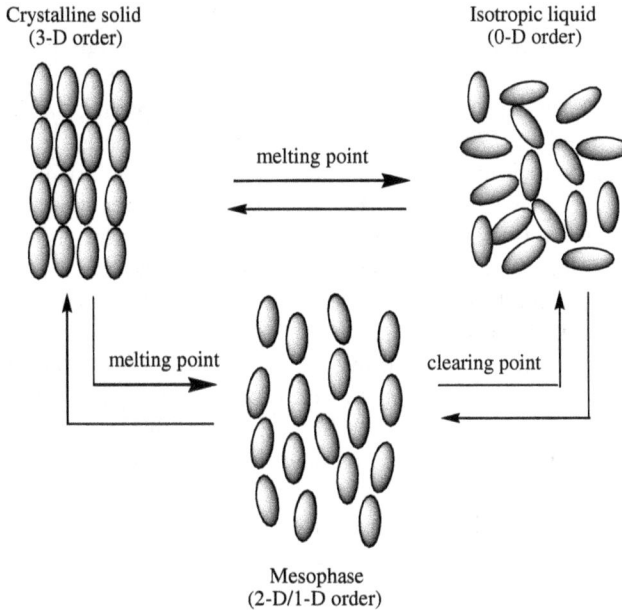

Figure 2 Melting behaviour of a liquid crystal.

impact and potential recycling opportunities. The chapter incorporates a review of recent literature, of which the following are particularly noteworthy and should be read in conjunction with the chapter:

i. *Desktop Computer Displays. A Life Cycle Assessment*, vol. 1, December 2001. EPA-744-R-01-004.
ii. Liquid Crystal Newsletter, No. 19, SID 2004 Special, Editor Dr Werner Becker, Merck KgaA, Darmstadt.
iii. E-Waste Curriculum Development Project. Phase 1: Research and Content Development – Literature Review, The Natural Edge Project, July 2006.
iv. 2008 Review of Directive 2002/96 on Waste Electronic and Electrical Equipment (WEEE): Final Report, United Nations University, Germany, August 2007.
v. Literature Review, Flat Panel Displays: End of Life Management Report, Final Report, King County Solid Waste Division, Seattle, USA, August 2007.

2 Overview of Liquid Crystals

Liquid crystals are unique substances that exhibit an intermediate state of matter that exists between its crystal and liquid states. Historically, the origin of liquid crystals dates back well over 100 years and is attributed to an Austrian botanist, F. Reinitzer, and a German physicist, O. Lehmann. In 1888, Reinitzer reported unusual melting behaviour in a sample of seemingly pure cholesteryl benzoate, which failed to give an expected clear melt at precisely 145.5 °C.[9] Instead, cholesteryl benzoate produced a persistently opaque, turbid liquid that eventually cleared at 178.5 °C. In 1889, Lehmann investigated this unusual *double-melting* behaviour using a specially constructed polarising light microscope and reported that the cloudy turbid fluid was in fact a new state of matter occurring between the liquid and solid states.[10] Lehmann is credited with introducing the terms *fliessende krystalle* and *flüssige krystalle* meaning 'flowing crystals', marking the birth of liquid crystals and onset of significant synthetic activity in search of more examples. Through extensive molecular structure–liquid-crystal property investigations, D. Vorländer formed the ground rules for liquid crystallinity and concluded that the basic molecular structure should be elongated, linear and rod- or lath-like.[11,12] This simple surmise is still used today in the design and synthesis of new liquid crystals for potential use in electro-optic LCD devices.

However, liquid crystals came to the fore in the late 1960s and early 1970s as academe and industry realised their huge commercial benefits rather than mere academic curiosity. The Radio Corporation of America (RCA) first demonstrated that a mixture of a liquid-crystalline phase and ionic dopant, sandwiched between two electrically conducting glass plates, produced an electro-optic scattering device (Dynamic Scattering Mode).[13] In the 'off' state (zero voltage applied), the liquid crystal was aligned so that the display appeared transparent. On application of an applied field (the 'on' state) the device appeared scattering, due to migration of ions to opposite electrodes causing electro-dynamic turbulence and re-orientation of the liquid crystal molecules. Unfortunately, the lack of sufficiently stable (chemical and electro-chemical) liquid crystals, the requirement for high driving-voltages and commercial pressures at the time rendered the technology redundant. The liquid-crystal materials used at the time were based on moisture-sensitive aromatic azomethines which were prone to hydrolysis and/or photolabile azoxy compounds. However, in 1971 Schadt and Helfrich[14] reported the Twisted Nematic (TN) display device and simultaneous intuitive developments in the synthesis of stable, room-temperature liquid crystals by Gray *et al.*[15] saw the real commercial birth of LCDs.

LCD technology has progressed rapidly since the development of the TN display and now large-area flat-panel LCDs, once a mere dream, are a reality in many households. Acronyms such as VA (Vertically Aligned),[16] IPS (In-Plane Switching)[16] and FFS (Fringe-Field Switching)[17] are all examples of new technology designed to improve viewing angles and enabling LCD TV.

2.1   Definition and Classification of Liquid Crystals

It is well known that matter exists in three forms: solid, liquid and gas. In a crystalline solid all the atoms, ions and molecules are fixed on regular lattice points and lattice planes, giving rise to three-dimensional order. As a crystalline solid melts it loses its three-dimensional order and transforms into a zero-order isotropic liquid state. Conversely, as an isotropic liquid is cooled, three-dimensional order is restored and the crystalline solid reforms (Figure 2).

Liquid crystals are a special class of compounds that neither convert to an isotropic liquid directly upon heating nor revert to a crystalline solid on cooling from the isotropic liquid state. Instead, they exhibit an intermediate state of matter (mesophase) possessing one- and two-dimensional order that exists between solid and liquid (Figure 2). Liquid crystals possess characteristics of both solids (anisotropic physical properties) and liquids (fluid-like), and some unique properties. The latter are exemplified by their unique ability to modulate light in response to an applied electric field, which is the basis for LCDs and their success today.

Liquid crystals capable of mesophase formation due to the action of heat, such as cholesteryl benzoate, are termed thermotropic liquid crystals. Alternatively, there are substances that generate a mesophase due to the action of solvents, usually water, which are termed lyotropic liquid crystals. Soap is an example of a lyotropic liquid crystal. The terms thermotropic and lyotropic are ambiguous, as lyotropic liquid crystals can also behave as thermotropic liquid crystals and *vice versa*.

Liquid crystals may be also classified on the basis of their chemical composition and solubility characteristics in organic and aqueous solvents. Non-amphiphilic liquid crystals are essentially hydrophobic, non-polar or moderately polar organic compounds that are soluble in organic solvents. Amphiphilic liquid crystals are compounds that contain both hydrophobic and hydrophilic entities in the chemical construct, conferring solubility in both aqueous and organic solvents. In the context of this chapter on electronic waste, lyotropic liquid crystals and amphiphilic liquid crystals will not be discussed further as they are not used in electro-optic LCD devices. The discussion will focus on thermotropic, non-amphiphilic, liquid crystals.

2.2 Molecular and Chemical Architecture of Liquid Crystals

Chemists play an important role in the design and synthesis of liquid crystals. Their ability to alter the chemical composition of a liquid crystal allows the development and understanding of important structure–property relationships necessary for future improved performance of LCD applications.

Liquid crystals used in electro-optic LCD devices are essentially non-polar or moderately polar small organic compounds composed of C, H, O and a halogen (usually F). The majority of known liquid-crystal compounds and those used in LCDs usually possess a geometrically anisotropic, elongated, rod- or ellipsoid-like molecular architecture (Figure 3) termed **calamitic**.

In order to maintain an elongated molecular geometry, a calamitic liquid crystal comprises terminal groups, **A** and **B**, which usually serve to lengthen the molecule whilst preserving linearity. The *rectangles* in Figure 3 represent rigid polarisable groups, usually aromatic ring systems, though cyclohexane rings

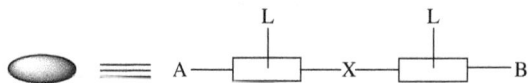

Figure 3   Schematic of a calamitic liquid crystal.

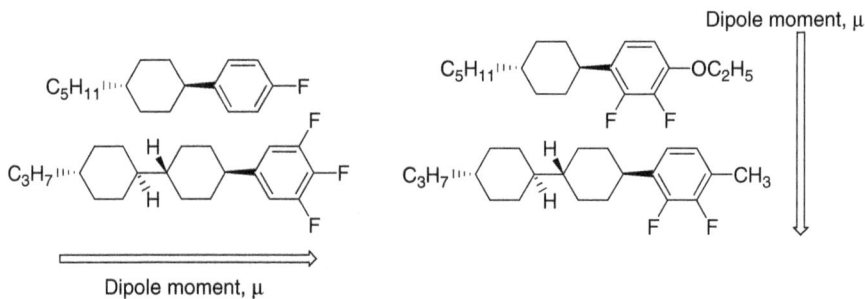

Figure 4   Examples of fluorinated liquid-crystals with positive and negative $\Delta\varepsilon$[18].

also are used widely. The central linking group, **X**, serves to maintain rigidity, linearity and, in some cases, may allow mesomeric relay of electrons between the two polarisable groups disposed on either side. Lateral substituents, **L**, serve to broaden a calamitic molecule and generally destabilise mesophase formation. However, lateral fluoro-substituents are used to good effect in the design of many new liquid crystals for commercial electro-optic LCD applications.[18,19] Lateral fluoro-substitution is important because it tends to lower melting point, enhance nematic-phase stability and, dependent upon the number and disposition of fluoro-substituents, enhance either positive or negative dielectric anisotropy, $\Delta\varepsilon$ (Figure 4).

Despite the relatively high cost associated with synthesis of fluorinated liquid crystals, most active-matrix LCDs contain fluorine, either as a part of a polar group or within the mesogenic core structure. Merck KGaA, Darmstadt, continues to lead the field and in 2003 was awarded the German 'Future Prize' for synthesis of super-fluorinated liquid crystals for use in Vertically Aligned (VA) technology.

2.3   The Mesophase: Types of Intermediate State of Matter

In any account of liquid crystals it is important to realise that there are three main types of intermediate state of matter or mesophase:[20] nematic, smectic and cholesteric.[i] Any liquid crystal will always exhibit at least one of the three types mentioned. The cholesteric mesophase only occurs in liquid crystals

---

[i] Cholesteric mesophase is also known as chiral nematic.

composed of chiral molecules (usually molecules possessing a carbon centre attached to four different groups). In the context of this chapter and in the nature of electronic LCD-waste, the nematic mesophase is most important since it is used in nearly all electro-optic applications. Nematic liquid crystals will be most abundant in LCDs containing WEEE. However, a brief description of all three mesophase types is given below because the smectic mesophase is used in high-specification digital cameras and the cholesteric mesophase (chiral nematic) is currently being evaluated in large-area billboard displays.

2.3.1 The Nematic Phase

The name nematic is derived from the Greek word 'nematos' meaning 'thread-like', because of its characteristic thread-like optical texture when viewed through a polarising light microscope. The molecules (shown as ellipsoids in Figure 5) in the nematic phase align approximately parallel with respect to each other, exhibiting weak long-range 1-D order. The molecules are free to rotate about both their long and short axes. The director, n, represents the direction of alignment and the extent of ordering is represented by the order parameter, S, which usually varies from 0.3–0.7. The nematic phase closely resembles the liquid state and is often said to be fluid-like.

2.3.2 The Smectic Phase

The name smectic is derived from the Greek word 'smectos', meaning 'soap-like'. The molecules in the smectic phase are more ordered than in the nematic phase. The molecules are stratified and occur in layers (Figure 6). Stratification may be either perpendicular (orthogonal arrangement) or tilted with respect to the layer planes. The molecules possess varying amounts of both orientation and positional order within and between layers. Depending on the type and extent, several smectic variants are possible and are classified as smectic polymorphs. The chiral smectic C-phase is an important smectic polymorph because it is used in surface-stabilised ferroelectric liquid-crystal (SSFLC)[21] devices found in small-area high-specification applications. The volume of

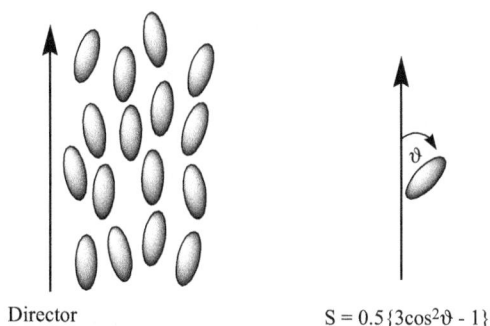

Director $S = 0.5\{3\cos^2\vartheta - 1\}$

Figure 5 Molecular arrangement of the nematic phase.

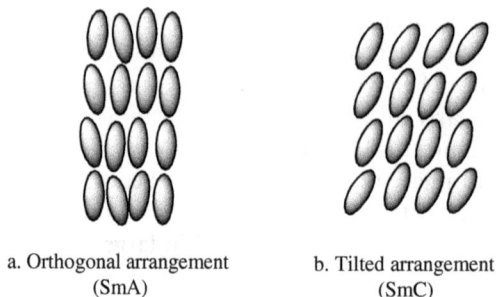

a. Orthogonal arrangement   b. Tilted arrangement
        (SmA)   (SmC)

Figure 6   Molecular arrangements of the smectic phase.

smectic waste in comparison to nematic waste will be insignificant and in the scope of this chapter neither smectic polymorphism nor SSFLC will be discussed further.

2.3.3   The Cholesteric or Chiral Nematic Phase
The name cholesteric is used for historical reasons, because this phase was routinely detected in derivatives of cholesterol. However, the cholesteric phase is exhibited by liquid crystals composed of chiral molecules irrespective of the presence of cholesterol and is now aptly termed the chiral nematic phase. The cholesteric or chiral nematic phase is envisaged as a pseudo-layered nematic state (Figure 7). Each layer comprises a nematic-like arrangement of molecules with a preferred orientation given by the director, *n*. Passing through several layers, the director skews uni-directionally, in either a clockwise or an anti-clockwise manner, depending on the 'handedness' of the molecules. A helical structure develops and the distance taken for the director to complete one full revolution (0 ° to 360 °) is termed the pitch length. The chiral nematic phase is used in reflective displays and temperature sensors because it has the ability to selectively reflect polarised light. Selectivity is dependent upon pitch length which itself is dependent upon temperature.

2.4   Physical Properties of Liquid Crystals and Material Requirements

The majority of electro-optic LCDs are based on the nematic phase. However, many practical considerations have to be satisfied by a liquid crystal prior to use in a display device:[16,22,23]

   i. The liquid crystal must exhibit a wide nematic temperature-range, usually −30 °C to +60 °C, allowing the device to work in both hot and cold climates.
  ii. The nematic must exhibit extreme chemical, electrochemical and physico-chemical stability. It should be air stable and not prone to moisture and/

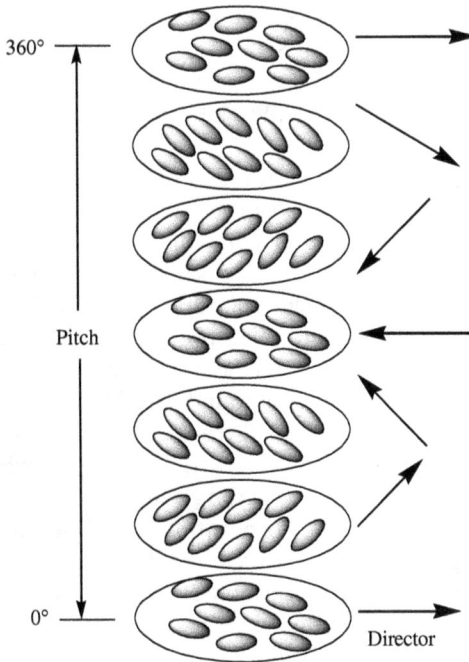

Figure 7 Molecular arrangement of the cholesteric phase.

or oxidative degradation, as was the case in the early displays based on azoxy- and imine- central-linking groups.

iii. The nematic phase must be responsive to an applied electric field, *i.e.* it must possess either positive or negative dielectric anisotropy ($\Delta\varepsilon$), defined as the difference between the dielectric constant parallel ($\varepsilon_{\parallel}$) and perpendicular ($\varepsilon_{\perp}$) to the nematic director. $\Delta\varepsilon$ determines the strength of interaction between an LC and an applied electric field and is proportional to the square of the molecular dipole moment, μ. The dielectric anisotropy affects the threshold voltage, V_{th}, of the electro-optical response and also the operating voltage of the driving circuitry of an LCD.

iv. LC molecules are switched in response to an applied electric field whose magnitude is related to the dielectric anisotropy. However, as the field is removed, the LC molecules revert to their original position due to inherent elastic forces. Elasticity in LCs is described in terms of curvature, expressed as splay, bend and twist deformations of the nematic director, termed k_{11}, k_{22} and k_{33} elastic constants, respectively. A low value of the ratio k_{33}/k_{11} is usually desirable for TN and VA mode.

v. The LC must possess low viscosity, allowing fast-moving images to be viewed. Viscosity is very complex to predict, as five independent viscosities are required to characterise a nematic liquid. In the context of this chapter it is sufficient to say the viscosity correlates well with response time, *i.e.* low viscosity infers short response time.

    vi. Material must possess low birefringence ($\Delta n = n_\parallel - n_\perp$) in order to give optimum viewing characteristics. The birefringence of a nematic liquid crystal is proportional to the order parameter S, but is also important in determining the optimal thickness of a display device.

Certain physical parameters ($\Delta\varepsilon$ and Δn) can be tuned by careful chemical design but viscosity and elastic constants are extremely difficult to correlate with respect to molecular structure. Unfortunately, to date there is no single material that produces a nematic phase satisfying all of the above requirements. Instead, a multi-component mixture of 15–20 different materials is blended to achieve the above-mentioned properties. The blend composition is specific to each LCD manufacturer and subject to strict confidentiality. In excess of 50 000 liquid crystal compounds are known; however, fewer than 1% are found in LCDs. There are three major suppliers of liquid crystals and liquid-crystal mixtures: Merck KGaA, Darmstadt, who are the largest supplier of liquid crystals with a very strong patent portfolio holding over 2500 patents, and their rivals Chisso Corporation and Dainippon Ink Corporation, both from Japan. In 2006, Merck reported impressive sales of liquid crystals worth 892 million euros, an increase of 21% over the previous year. Merck have successfully penetrated the Far East with the successful opening of Merck Advanced Technology Ltd., Korea, in 2002 and commissioning of a new liquid-crystal centre in Taiwan to meet local rising demand in 2006. Although not in the same power-base as Merck, Chisso and Dainippon Ink, many new Chinese and Taiwanese liquid-crystal chemical manufacturers are emerging.

3 Overview of Liquid Crystal Displays Based on Nematic Mesophase

3.1 Basic LCD Operating Principles

Electro-optic LCDs may be regarded as optical shutters that either transmit or block polarised light in response to an applied electric field. A minimum threshold voltage must be reached before the shutter can work, based on re-orientation of the liquid-crystal molecules. If an applied voltage is below the threshold level, then no re-orientation occurs, but exceeding this level causes the liquid-crystal molecules to re-orient either with the field (parallel) or opposed (perpendicular). An LCD is made of many small segments or picture elements (pixels). Each pixel contains electrically addressable liquid crystal. High-information-content LCD devices require many thousands of pixels built as a 2-D array of N rows and M columns giving a total of ($N \times M$) pixels. Each pixel may be addressed (driven) individually, requiring a total of ($N \times M$) electrical connections. This situation, known as direct addressing, is not feasible for high-resolution displays but is found in low-resolution displays (> 50 pixels) such as watches, timers and calculators. For medium and higher resolution

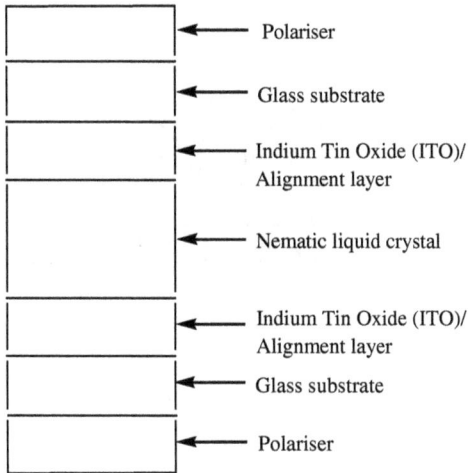

← Polariser

← Glass substrate

← Indium Tin Oxide (ITO)/
Alignment layer

← Nematic liquid crystal

← Indium Tin Oxide (ITO)/
Alignment layer

← Glass substrate

← Polariser

Figure 8 Basic construction of an electro-optic LCD device (side view).

laptop monitors and LCD TVs, the number of connections is reduced to $(N + M)$ using the concept of active-matrix addressing (multiplexing). Every pixel is addressed by the cross-over region between a row and a column. Active-matrix displays are dominant and are usually referred to as AMTFT-LCD. A TFT is a combined small transistor and capacitor unit located at every pixel, acting as a variable on and off switch.[16]

The basic construction of an electro-optical LCD device shown in Figure 8 comprises a thin film of nematic mesophase sandwiched between two glass substrates whose inner surfaces have been coated with a transparent electrical conducting layer of indium tin oxide (ITO), allowing for a voltage to pass across the medium. Dependent on the display mode, the nematic mesophase is pre-aligned, aided by preferentially rubbed alignment layers (polyimide, approximately 100 nm thickness), which are also coated on the inner surfaces of the glass. The two glass substrates are kept apart by a few microns using spacer beads or spacer posts. Polarisers are attached to the outer surface of the glass substrates and their arrangement dictates the appearance of the display, either dark or bright in the 'off' state. The role of the nematic mesophase is two-fold: (i) to modulate incoming polarised light and (ii) to respond to an electrical impulse by aligning with the applied field.

3.2 Types of Electro-optic LCD Devices

In the scope of this chapter an awareness of the mode of action of the three main types of electro-optic display device (see Figure 9) utilising the nematic mesophase, which are to be expected in the waste stream, is essential: Twisted Nematic (TN)[14] (including STN, TFT-TN), In-Plane Switching (IPS)[16] and Fringe-Field Switching (FFS),[17] and Vertically-Aligned Nematic (VAN).[16] The

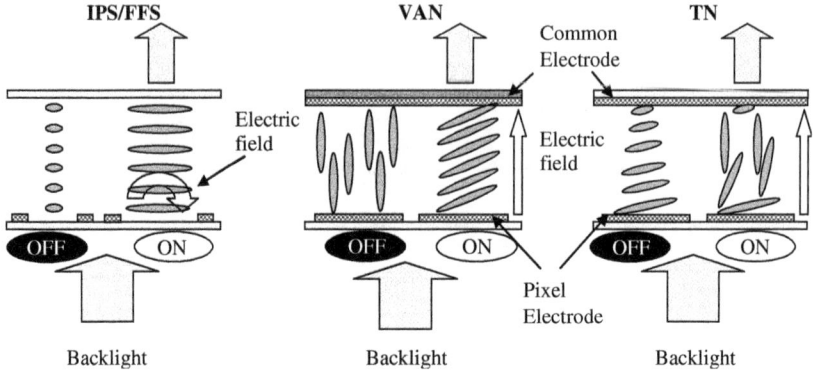

Figure 9 IPS/FFS, VAN and TN display modes.

mode of action determines the nature of the liquid-crystal mixture used in each display type. TN and IPS technologies require liquid-crystal mixtures with positive dielectric anisotropy, $+\Delta\varepsilon$, whereas VAN technology requires materials with negative dielectric anisotropy, $-\Delta\varepsilon$.

For purposes of re-deployment in electro-optic applications, care must be taken not to mix TN and IPS/FFS LCD panels with VAN LCD panels. Additionally, there are subtle variations of each display type that improve contrast and resolution, but these are dependent on subsidiary factors such as quality of backlight, brightness-enhancement films (BEF), dual brightness-enhancement films (DBEF), ITO electrode patterning and geometry, polarisers, anti-glare coatings and anti-reflection films.

3.2.1   The Twisted Nematic Mode (TN)

The Twisted Nematic (TN) Display, developed by Schadt and Helfrich[14] in 1971, is perhaps the most common LCD display type. A TN-LCD (see Figure 10) consists of a thin film of homogeneously aligned (the molecular long-axis of the molecules oriented parallel with respect to the glass substrates) nematic liquid-crystals sandwiched between two glass plates, separated by a distance of 3–10 µm. The nematic liquid crystal is twisted by 90° between the two glass plates. The innermost surfaces of the glass are coated with transparent, electrically conducting indium-tin oxide (ITO), on top of which a polyimide alignment layer (approximately 100 nm thickness) is deposited; this controls the orientation and alignment of the liquid crystal molecules in the 'off' state (zero voltage). The polyimide layer is rubbed in a uniform direction, but between the two glass plates the rubbing direction is offset by 90°. The alignment layer forces the nematic liquid crystal molecules to align parallel with each inner surface (homogeneous alignment), but because of the offset rubbing direction the molecules are twisted through 90° on passing from one surface to the next, hence the name twisted nematic. To ensure a homogenous twist structure a

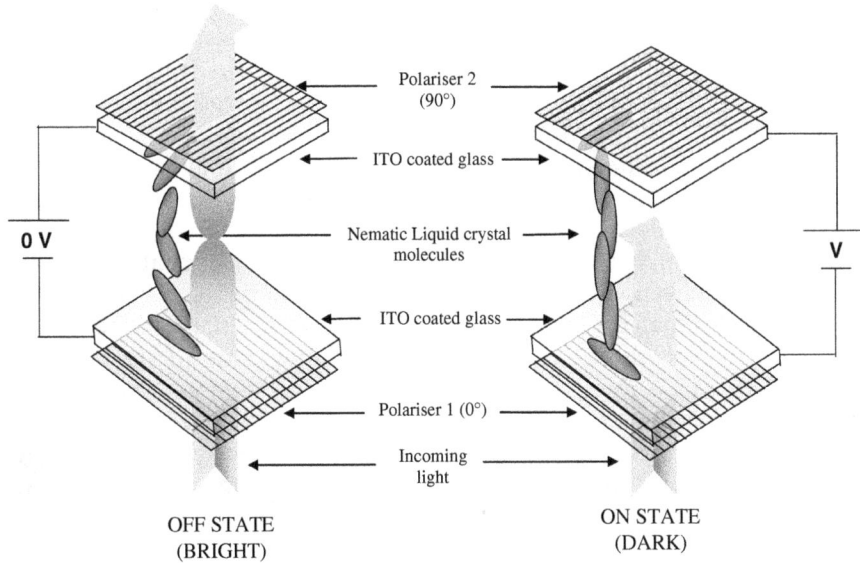

Figure 10 Operation of TN mode.

small amount of chiral dopant (up to 0.1%) may be added to the liquid-crystal mixture. The outer surfaces of the glass are laminated with polarisers, with their transmission direction parallel to the rubbing direction of the polyimide layer.

Non-polarised light enters the bottom polariser and becomes linearly plane polarised, vibrating in the direction of the longitudinal axis of the liquid-crystal molecules. As it passes through the twisted nematic liquid-crystal, it is said to be wave-guided such that the plane of polarised light is rotated by 90° as it reaches the top polariser. The polarisation plane of the light matches the polarisation plane of the polariser and is able to exit, producing a bright or normally white (transmissive 'off'-state) field of view. In the case of colour displays, light passes through red, green and blue colour filters before reaching the viewer.

On application of an electric field, the twist arrangement of the nematic liquid-crystal molecules is destroyed as the molecules re-orient parallel with respect to an applied field and perpendicular (homeotropic alignment) with respect to the glass substrates. Plane polarised light entering the liquid crystal medium is no longer rotated and on reaching the top polariser it is absorbed, giving a dark (black) field of view in the 'on'-state. The twisted arrangement reforms when the field is removed and the display is ready to be addressed again. LCD watches or calculators show a silver-grey appearance in the 'off'-state because a reflective-mirror strip is attached on the bottom glass plate below the lower polariser. Ambient incoming light is reflected by the mirror-like strip and emerges back out of the top polariser.

3.2.2   In-Plane Switching (IPS) and Fringe-Field Switching (FFS) Modes
In-Plane Switching[16] was developed to counteract the problem of poor viewing
angle commonly associated with TN-LCD, which relies on an out-of-plane re-
orientation of the liquid-crystal molecules (planar to perpendicular and *vice
versa*). IPS technology is found in large-area LCD TVs. In IPS-LCD the
molecules are modulated in the same plane in response to an applied electric
field, hence the name in-plane switching. To create this effect both electrodes
are patterned on the same glass surface and not on both as in TN-LCD.
The nematic crystal molecules adopt a planar-parallel arrangement with respect
to the rubbing direction of the polyimide layer. The polarisers are set crossed,
i.e. one polariser matches the rubbing direction whilst the other is offset by 90°.
In the absence of an electric field, the polarisation state of the incoming
polarised light remains unchanged and is absorbed by the second polariser,
giving a normally black state. Application of an in-plane electric field 45° to the
rubbing direction causes the molecules to align in-plane with the applied field.
A pseudo-twist structure is formed, capable of changing the polarisation state
of the incoming polarised light such that it matches the second polariser and
can exit.

   Although a wide viewing-angle is a major selling point of IPS, it has to
overcome, and in some cases has done so successfully, issues of smaller pixel
aperture, leading to brighter backlighting and complex power-hungry drive
electronics.

   The basic principle of FFS[17] is quite similar for AS-IPS (Advanced Super-
IPS). The key feature is a non-patterned (no slits) common electrode situated
directly under the pixel electrode. Electrical fringes of higher efficiency are
achieved, thus improving brightness and contrast ratio. FFS technology offers
a contrast ratio as high as 1:700 and wide viewing-angle close to 180°. More-
over, its driving voltage is roughly 4 V, much lower than 10 V of multi-domain
vertically-aligned (MVA) and IPS, making it one of the least power-consuming
panels.

3.2.3   Vertically Aligned Mode (VA)
The VA mode also boasts wide viewing and high contrast as its key selling point
and is in direct competition with IPS. The VA mode is also found in large area
LCD TVs. The VA mode is similar to the TN mode, except that the nematic
liquid-crystal molecules are aligned vertically (homeotropic alignment) and
possess a negative dielectric anisotropy ($\Delta\varepsilon$).[16] A consequence of the latter is
that the molecules will oscillate (re-orient) perpendicular with respect to an
applied field. In the 'off'-state a non-transmissive, normally black state is
observed as polarised incident light from the first polariser is absorbed by the
second polariser. Application of an electric field re-orients the molecules such
that they align at 45° with respect to both polarisers, producing a transmissive
'on' state. Holding the molecules at 45° requires special surface treatments and
surface architectures, which has led to several variations of VA technology,

namely Patterned and Multi-domain Vertically-Aligned, PVA[16] and MVA[16], respectively.

4 LCD Manufacturing Process

The manufacturing process is a highly expensive automated process predominantly undertaken in Japan, Korea and China (both mainland and Taiwan). The success or status of an LCD-manufacturing facility is based on the maximum glass size it can process. An economy-of-scale principle applies as more panels can be fabricated from a single larger pane of mother glass, involving fewer processing steps. For example, the first fabrication plants (Generation 1) came on stream in 1990, capable of handling 300×400 mm glass panels from which a single 15-inch display could be manufactured. The rapid success of LCD and under-pinning technologies, coupled with a dramatic demand for large-area LCD TV, now sees the emergence of state-of-the-art Generation 8 fabrication facilities capable of processing 2160×2460 mm glass panels, from which eight 46-inch wide panels can be cut. The industry continues to progress and Sharp, in alliance with Sony, recently indicated plans to build the world's first 10th generation (2850×3050 mm) facility by 2010, capable of yielding six 65-inch panels, nine 50-inch panels or fifteen 40-inch panels. The manufacturing process differs slightly from manufacturer to manufacturer, based on their product line and budget. Figures 11–14 show the basic processes for TFT-TN panel fabrication which are:

i. The Array Process (Figure 11) involves pre-treatment and deposition of layers of the inner surface of the front colour filter and the rear TFT glass:

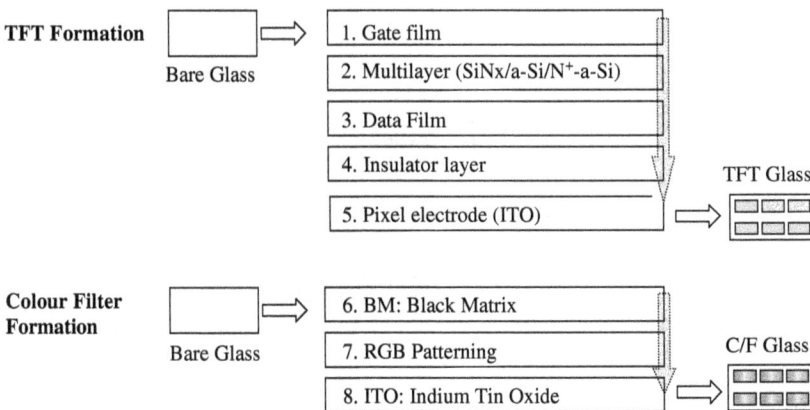

Figure 11   The array process.

A Film formation Bare Glass →

| Film Deposition |
| Photolithography |
| Etching |
| Stripping |

Glass with film →

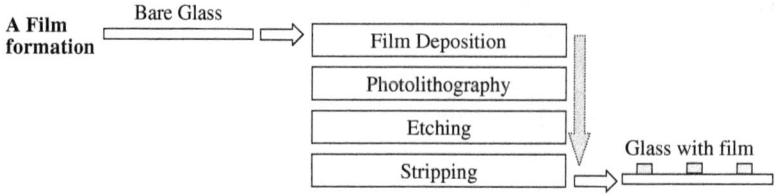

Figure 12 Protocol for film formation.

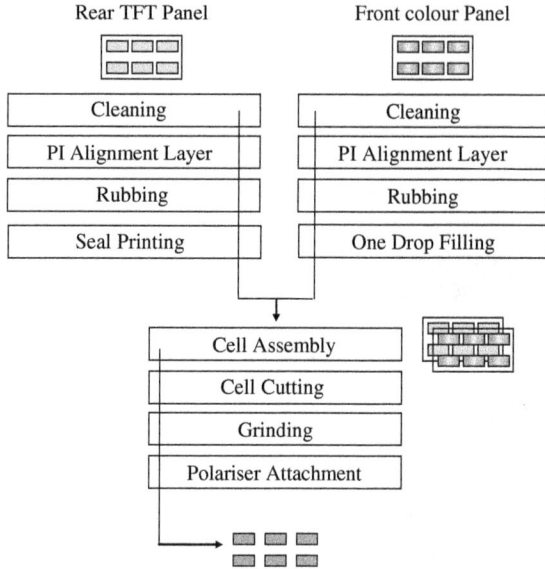

Rear TFT Panel   Front colour Panel

Cleaning		Cleaning
PI Alignment Layer		PI Alignment Layer
Rubbing		Rubbing
Seal Printing		One Drop Filling

| Cell Assembly |
| Cell Cutting |
| Grinding |
| Polariser Attachment |

Figure 13 The cell process.

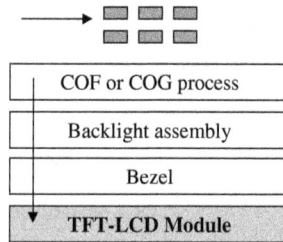

| COF or COG process |
| Backlight assembly |
| Bezel |
| **TFT-LCD Module** |

Figure 14 The module assembly process.

Steps 1–7 shown in Figure 11 are produced using photolithographic processes as depicted in Figure 12. Step 8 only involves deposition of ITO without patterning.

ii. The Cell Process (Figure 13) joins front and rear panels which are then filled with liquid crystal.

One-Drop Filling (ODF) is a new technology used to dispense liquid crystals, replacing conventional vacuum-capillary filling. The liquid crystal is dispensed drop-wise, hence the name one-drop filling. ODF technology is highly beneficial because it uses less liquid-crystal and saves considerable time. For example, a 30-inch panel requires approximately 5 days to fill using conventional filling, whereas ODF takes only 5 minutes.

    iii. The Module Assembly Process (Figure 14) involves connecting additional components such as driver-integrated circuits and backlights.

The processes outlined in Figures 11–14 are only a small sub-set in the manufacturing process of an LCD device such as a computer monitor. The LCD module has to be fitted with power supply, sound box, driver circuitry and frame to give the final product that we recognise as an LCD monitor or TV (Figure 15).

5 Environmental Legislation and Lifecycle Analysis

This section should be read in conjunction with respect to references 24–27 as they represent the current, most comprehensive reviews on environmental legislation and lifecycle analysis of LCDs.

An LCD is a complex mix of glass, plastic, metal (some of which are precious, *e.g.* indium), electronic circuitry and the liquid-crystal itself. Every aspect of an LCD device is now under scrutiny from a series of environmental regulations and measures that makes manufacturers accountable for their process not only from cradle but also, more importantly, to grave (end of life, EOL). More recent regulations probably stem from the Basel Convention,[3] formulated in 1992 to prevent free transport of toxic, hazardous waste from rich Nations (OECD – Organisation for Economic Co-operation and Development) to their lesser-developed poor neighbours, even under the guise of recycling. One hundred and seventy nations have ratified the Basel Convention but, notably, the United States has yet to ratify. However, the US Environmental Protection Agency (US EPA) has made significant effort to evaluate and understand the nature of LCDs (and CRTs) by commissioning studies on their Lifecycle Analysis (LCA) and potential EOL implications.[24]

There are many environmental regulations, global, national, regional and local, of which WEEE, RoHS and REACH probably are the most relevant to LCD waste.

5.1 The WEEE Directive and LCDs

The Waste Electric and Electronic Equipment Directive[4] (2002/96/EC) was adopted by the EU on January 2003 and recommended that Member States should convert the directive into national law by 2004. The WEEE Directive aims to divert the amount of electronic waste entering landfill or being incinerated by encouraging recovery and re-use. Electronic equipment is divided in

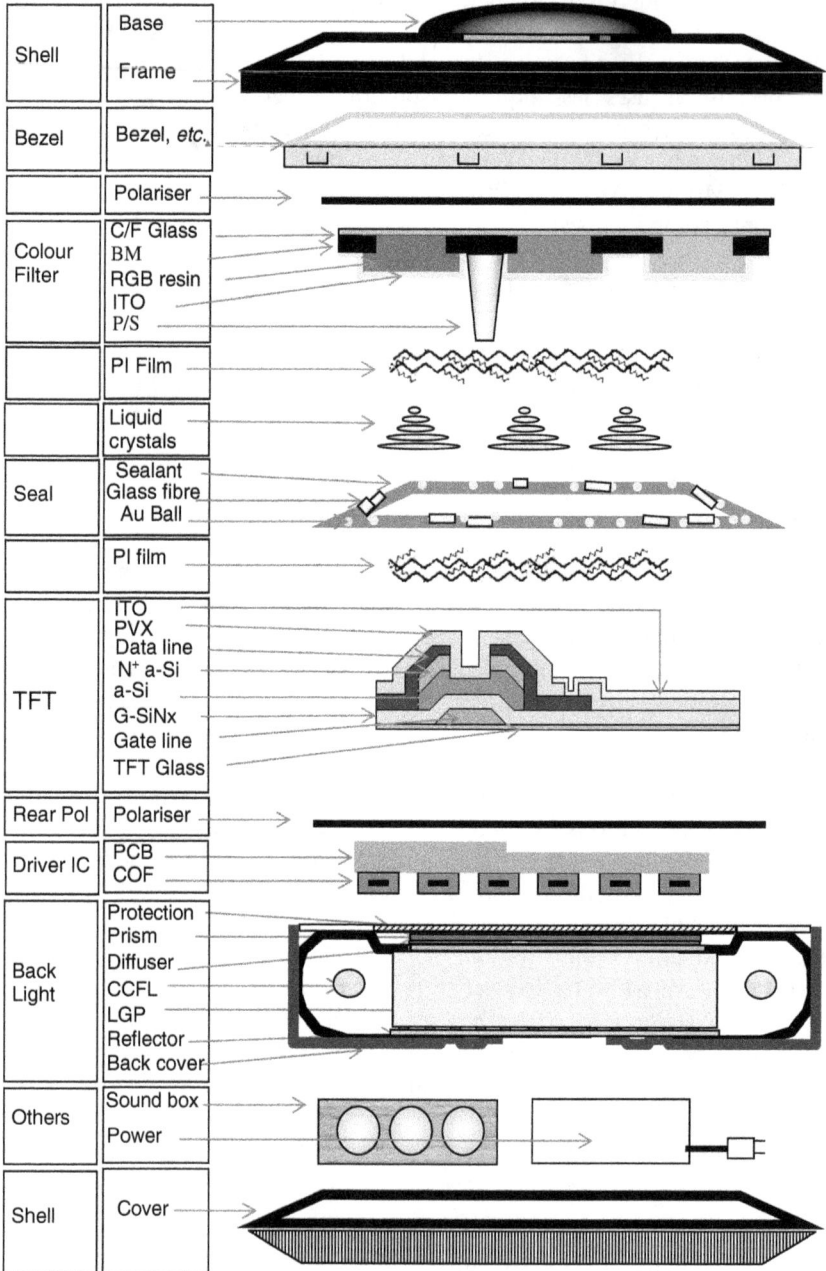

Figure 15 Complete assembly of an LCD device.

to ten broad categories, of which Categories 3C (LCD monitors), 4C (flat panel TVs) and 5B (lighting equipment) are pertinent to LCDs. The WEEE Directive (2002/96/EC) stipulates that LCD-containing WEEE with a surface area greater than $100\,cm^2$ and those containing mercury backlights must be isolated from the waste stream. Recovery and re-use/recycling rates of 75% and 65%, respectively, have been set for computer-related equipment.

5.2   RoHS and REACH

LCDs are also governed by regulations complementary to WEEE. The RoHS (Restriction of Hazardous Substances, 2002/95/EC, implemented in UK July 2006[5]) Directive restricts use of lead, mercury, cadmium or hexavalent chromium, and brominated flame retardants (PBDE and PBB) in electronic goods. REACH (Registration, Evaluation, Authorisation and Restriction of Chemicals, implemented UK April 2007[6]) ensures a high level of protection of human health and the environment as well as the free circulation of substances on the internal market while enhancing competitiveness and innovation.

5.3   Far East Environmental Measures

As Europe and the US are major consumers of LCD devices and have introduced strict environmental legislation, the Far East and Asian LCD-producing countries (Korea, Japan and China, including both mainland and Taiwan) have had to introduce new laws comparable to WEEE, RoHS and REACH. In Japan, the Green Procurement Law, the Energy Conservation Law, RoHS Legislation (Class 1 and Class 2 substances), Product Take-back legislation (*e.g.* WEEE) and Prefecture local environmental laws need to be considered and enforced. In China, the Clean Production Promotion Law, State Environmental Protection Standards, China RoHS (February 2007) and Energy Conservation Law (2008) are coming into force. The Chinese Government is drafting Product Take-Back regulations but, surprisingly, eco-labelling is voluntary.[28]

5.4   Lifecycle Analysis

The US Environmental Protection Agency (USEPA) in 2001 published data on the lifecycle assessment of desktop computer displays, focusing on LCD *versus* CRT monitors.[24] Although many of the manufacturing protocols have now changed in the light of environmental regulations as the LCD manufacturing industry adopts greener solutions, this publication still serves as an excellent insight of the pros and cons of LCDs in the environment. The environmental and health impacts of LCDs were based on three key impact categories: (i) natural resource impacts, (ii) abiotic ecosystem impacts and (iii) human health and ecotoxicity. Each category was further sub-divided, giving a total of

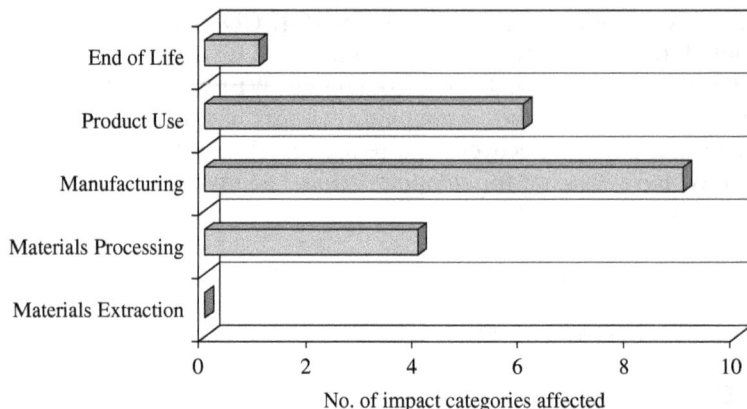

Figure 16 Lifecycle analysis of LCD computer monitors and associated impact categories.

20 impact categories. LCD monitors were found to have lower environmental impacts in 18 of the 20 impact categories studied.

Comparison of the different stages of an LCD life-analysis revealed that manufacturing and product use contributed to the greatest number of impact categories,[24] as shown in Figure 16.

The manufacturing process is very energy-intensive and involves chemical treatment, thus impacting mainly on abiotic aquatic systems and on human health and toxicity. At the time of this study, materials extraction revealed zero impact; however, it may have been subsumed in the manufacturing process.

The report also focused on the key question of toxicity of lead, mercury and liquid-crystals themselves. Lead is a major component in CRT glass whereas LCDs use lead-free glass. The reported concluded that CRTs have over 25 times more lead than LCDs. Mercury is highly toxic and may be perceived as a greater problem in LCDs than CRTs as it is contained in the cold-cathode fluorescent lamp (CCFL) backlight-unit in the amount of 2–7 mg of mercury. However, mercury is also emitted from some fuel combustion processes, such as coal-fired power stations, that contribute to lifecycle impacts of both CRTs and LCDs. The study found that more mercury was emitted from the generation of power consumed by CRTs during manufacture and use (7.75 mg) than from the entire amount of mercury emissions from the LCD, including mercury from backlights and power generation (7.21 mg). Since mercury is deliberately used in LCD manufacturing, mercury in LCDs has a wider effect on the environment that mercury from CRTs. LCD manufacturers are concerned both with cost, as the backlight unit forms a significant part of the bill of materials, and environmental legislation. There is a strong drive to move away from CCFL backlighting to light emitting diodes (LEDs) and external electron fluorescent lamps (EEFLs), especially for the large-area LCD TV market.

6 Potentially Hazardous Constituents: Toxicity of LCD Constituents

There are many components associated with an LCD device made from many materials, as listed in Table 1.[27] Potentially hazardous materials include chromium, found in the black matrix during colour-filter fabrication. Chromium is now being replaced by polymeric resins. Solder joints contain lead, but now there is a move towards lead-free solder. Small quantities of arsenic (As_2O_5) are used in glass manufacture. Corning have recently announced EAGLE XG™ glass, which is free of any heavy metal and arsenic. Annex II of the WEEE Directive attempts to de-pollute WEEE by ensuring effective removal and treatment of hazardous components. Perhaps the two most contentious materials are mercury, used in the CCFL lamps as part of the backlight assembly, and the liquid-crystal mixture itself, which is a chemical cocktail of up to 20 different organic compounds, comprising mainly the elements C, H, N, O and F.

6.1 Toxicity of Mercury and Backlighting

Mercury is found in most LCDs as part of the CCFL backlight unit. A CCFL is a sealed glass tube, filled with argon (or neon) and mercury (2–7 mg), with electrodes on both ends. The inner surface of a CCFL glass tube is coated with a phosphor and the wavelength, or emission colour, of light depends on the type of the gas and appropriate mix of red, green and blue phosphors. When high voltage is applied to the electrodes, ultraviolet energy at 254 nm is produced as the mercury and the internal gases are ionised. The resulting ultraviolet energy from the mercury discharge stimulates the phosphor lining inside the lamp, producing visible light in the wavelength region 380–780nm (also known as the photopic region).

The amount of mercury in a large-area LCD TV containing many CCFLs may exceed 300 mg. Mercury is classified as hazardous and is registered on the Hazardous Waste List (HWL) and on the European lists of wastes with codes 16 02 13 and 20 01 21.

LCD manufacturers are looking to reduce the number of CCFLs used or adopting alternative technologies. U-Shaped CCFL lamps are being employed by some manufacturers in 32-inch LCD TVs, which reduces the number of CCFLs by 50%, thereby lowering costs. Light-emitting diodes (LEDs) are now increasingly being used in 32, 37 and 40-inch LCD TVs as they are brighter, do not emit as much heat, use less space and are mercury-free. External-electrode fluorescent lamps (EEFLs), which, as the name implies, have electrodes on the outside, are being adopted by LG-Phillips. EEFLs are aimed at reducing energy costs because a single inverter is needed to drive several lamps, unlike CCFL technology that needs as many inverters as CCFLs used to achieve lamp-to-lamp uniformity. However, EEFLs still contain mercury as light emission operates on a similar principle to CCFL. Other technologies include Flat Fluorescent Lamps (FFL) and Hot-Cathode Fluorescent Lighting (HCFL),

Table 1 Types of materials found in an LCD device.

Material	Component	Part
Aluminised mylar	Backlight assembly	Corner tape
Aluminium	LCD assembly; power supply assembly	Glass panel assembly (thin film transistor); heat sink
Beryllium copper	LCD assembly; rear cover assembly	Metal clip, beryllium-copper fingers
Borosilicate	LCD assembly	Glass panel assembly
Brass	Backlight assembly	Brass threaded stand off
Chromium	LCD assembly	Glass panel assembly (black matrix for colour filter)
Discotic liquid crystal	LCD assembly	Glass panel assembly (compensation films)
Epoxy resin	LCD assembly	Glass panel assembly (sealant)
Foam rubber	Backlight assembly	Gasket
Gold	LCD assembly	Glass panel assembly (electrical conducting dots)
Glass	LCD assembly; backlight assembly	Glass panel assembly; light assembly (CCFL)
Hi-mu ferric	Backlight assembly	Flat cable toroid
Indium-tin oxide (ITO)	LCD assembly	Glass panel assembly (electrode)
Iodine	LCD assembly	Glass panel assembly (polarisers)
Lead	LCD assembly	Solder
Liquid crystals	LCD assembly	Glass panel assembly
Mercury	Backlight assembly	Light assembly (CCFL)
Molybdenum	LCD assembly	Glass panel assembly (thin film transistor)
Nylon	Backlight assembly; base/stand assembly	Nylon clamp; strain relief bushing
Phosphors	Backlight assembly	Light assembly (CCFL)
Plastics and plasticisers (phthalates)	Power supply assembly; rear cover assembly; base stand assembly	Power cord receptacle; rear covers
Polycarbonate	Backlight assembly	Light guide; rear plate assembly
Polycarbonate, glass-filled	LCD assembly	Plastic frame
Polyester	LCD assembly; power supply assembly; backlight assembly; rear cover assembly	Brightness enhancement films; opaque diffuser; white reflector; insulator; power switch
Polyester, glass-filled	Base/stand assembly	Upright
Polyimide	LCD assembly	Glass panel assembly (alignment layer)
Polymethyl methacrylate	Backlight assembly	Clear protector
Polyvinyl alcohol (PVA)	LCD assembly	Glass panel assembly (polarisers)
Resins	LCD assembly	Glass panel assembly (colour filter)

Table 1 (*Continued*).

Material	Component	Part
Silicon nitride (SiNx)	LCD assembly	Glass panel assembly (thin film transistor)
Silicone rubber	LCD assembly; back-light assembly; base/stand assembly	Gaskets; light assembly (shock cushion); rubber feet
Silver	LCD assembly	Glass panel assembly (electrical conducting dots)
Soda lime	LCD assembly	Glass panel assembly (glass)
Stainless steel	Base/stand assembly	Swivel bearing
Steel (iron)	Power supply assembly; backlight assembly; LCD assembly; rear cover assembly; base/stand assembly	Housing; screws; metal plate; rear plate; hold-down plate; metal plate brackets; washers; axle and spring; base weight; C-clip
Triacetyl cellulose (TAC)	LCD assembly	Glass panel assembly (polarisers)

Table 2 Liquid-crystal mixture content in LCD devices.

LCD Device	Area (cm^2)	LC (mg)	LCD panel weight (g)	LC/LCD panel (% by weight)
15-inch monitor	700	350.0	320	0.11
17-inch monitor	893	446.5	400	0.11
30-inch TV	2787	1193.5	1110	0.11

but these are still at too early a stage, even though FFL is expected to be cheaper than any other light sources for LCD TV.[29]

6.2 Toxicity of Liquid-crystal Mixture[7]

An LCD device typically contains 0.5 mg liquid-crystal mixture per cm^2 of display area. Table 2 lists the amount of liquid-crystal material contained within 15- and 17-inch LCD monitors and a 30-inch LCD TV. The latter contains approximately 1.2 g of liquid-crystal mixture. This may seem an insignificant amount except when placed in context of sales and shipments of 30-inch LCD TVs, for which a staggering 93 tonnes of liquid-crystal mixture was produced in 2007 (based on 78.2 million units sold or shipped in 2007).

Based on toxicological and ecotoxicological studies conducted by Merck KGaA, Darmstadt, the world's largest supplier of liquid-crystal mixture, the liquid crystals used in modern-day displays are classified as non-hazardous and coded 16 02 16 (provided that the mercury backlight has been removed). Landfill of LCDs is legally a 'Disposal' but contradicts the WEEE targets.

The LC manufacturers Merck, Chisso and Dainippon Ink Corporation agreed not to market any acutely toxic or mutagenic liquid-crystals. However, long-term exposure data and data concerning accumulation of LC following end of life are scant. LCs in the environment should be further investigated.

Research by Merck reveals:[7]

  i. Liquid crystals are not acutely toxic.
  ii. A few can be irritant, corrosive or sensitising, but these effects can be avoided by limiting the concentration of these substances in mixtures.
  iii. Liquid crystals are not mutagenic in bacteria (Ames test) or in mammalian cells (MLA screening; test with mouse lymphoma cells).
  iv. Liquid crystals are not suspected of being carcinogenic.
  v. They are not harmful to aquatic organisms (bacteria, algae, daphnia or fish). Ten individual liquid-crystal compounds and one representative mixture were tested. Concentrations of liquid crystals in the test media were measured using high-performance liquid chromatography (HPLC); toxicity tests were performed according to OECD and EU guidelines. Results for daphnia immobilisation and algal growth inhibition showed no adverse effects to aquatic organisms up to the limit of solubility of the liquid-crystal compounds. Based on these results, the German Federal Environment Agency does not enforce special requirements for the disposal of LCDs based on their content of liquid crystals. Recent tests by Merck include 6 for acute toxicity in fish, 8 for bacterial toxicity, 12 for algae growth inhibition and 36 for daphnia immobilisation, all of which reportedly show no hazard to aquatic organisms.
  vi. Some are not readily biodegradable.
  vii. In LCDs the amount of liquid-crystals is, at maximum, 0.1% by weight relative to the collected WEEE with area greater than 100 cm^2.

Despite extensive studies undertaken by Merck showing that LC material is essentially non-hazardous, they do recommend incineration rather than landfill. However, special incineration plants are required which maintain an elevated temperature throughout.

In Germany, if a decision is needed whether to deposit waste in a landfill designated for municipal or hazardous waste, then a chemical analysis is needed. Merck have undertaken such studies on a variety of LCDs filled with Merck liquid-crystals. Correlating the results of the analysis of LCDs with the criteria fixed under German waste regulation, it can be shown that isolated LCDs can be disposed of in landfills designated for municipal waste.

6.3   Demanufacture and Recycling

A possible protocol for collection and recycling of an LCD device is shown in Figure 17.

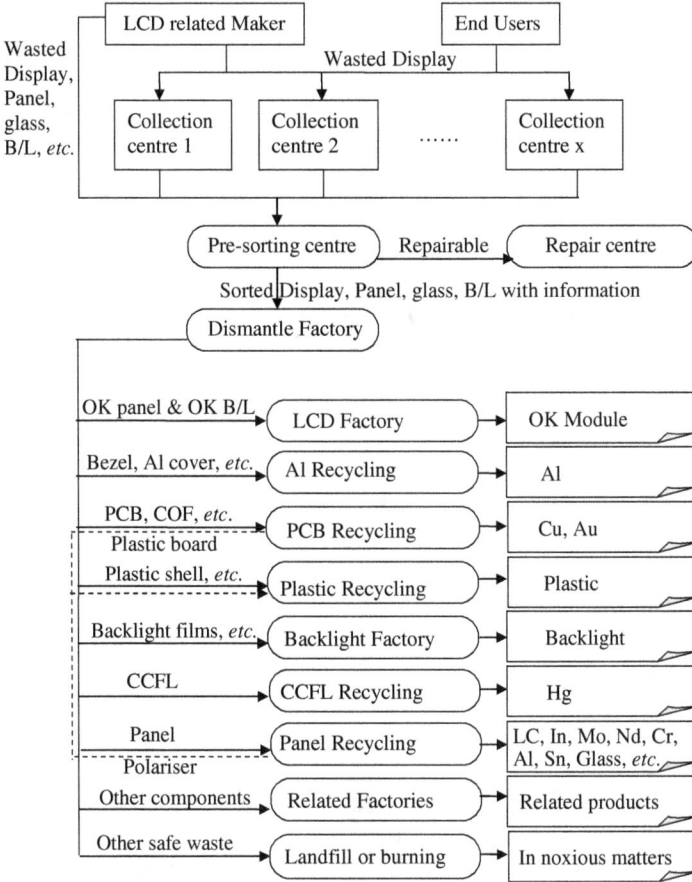

Figure 17 Collection and de-manufacturing pathway.

A TFT-LCD device comprises many sub-components as shown in Figure 18. The LCD module can be further disaggregated into two main components: the backlight unit and the LCD panel. The latter comprises colour filter, sealant, liquid-crystal mixture, glass and TFT array.

Glass used in TFT-LCD panels has a high melting point (approximately 1150 °C) and comprises 50% SiO_2, 15% B_2O, 10% Al_2O_3 and 25% BaO.[7] Since glass is a major component of an LCD panel (approximately 90% by weight), and with large volumes of glass produced annually for the TFT-LCD industry, it is feasible to consider its re-use or diversion from landfill. Providing that the mercury backlight is removed from an LCD panel, then the remaining constituents, including the liquid-crystal mixture, are classified as non-hazardous. Merck report the use of LCD glass instead of silica to line the inner walls of incinerators, thus protecting them from aggressive substances.[7]

The German recycling company Vicor GmbH, Berlin, separates the liquid crystal from glass by suction. The resultant liquid crystal is incinerated at

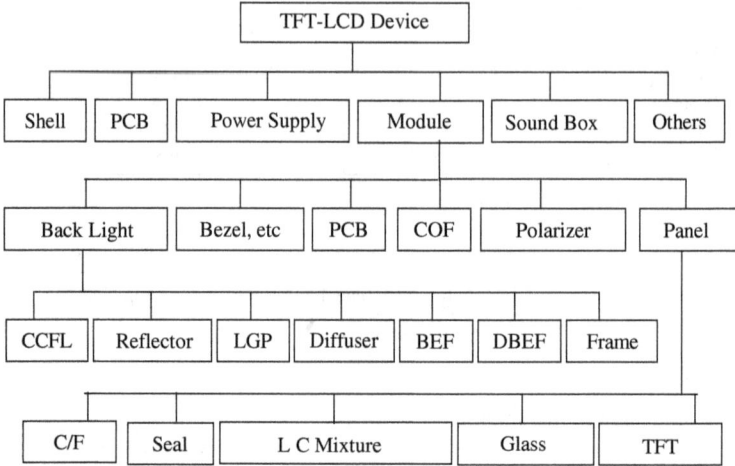

Figure 18 TFT-LCD device disassembly.

high temperature catalytically and up to 70% of the glass is recycled from the display.[27]

Removal of liquid crystals by catalytic decomposition is well documented in the literature. For example, Sharp Corporation report decomposition of liquid crystals using photocatalysts (*e.g.* TiO_2). The decomposition uses low energy and does not require waste-water treatment.[30]

Many solutions for recovering LCs rather than incinerating them are now being disclosed in patents. Dainippon Ink & Chemicals (DIC)[31,32] have patented a method for removal and purification of liquid-crystal components from waste liquid-crystal such that the isolated components can then be re-used as additives in mixtures used for active-matrix displays. Isolation and purification is achieved by column chromatography and followed by distillation. The isolated components are most likely ester- and cyano-group-containing liquid crystals.

Sumoge and Hiruma[33] studied the thermal decomposition, gasification and melting of LCD-containing WEEE, allowing for a process of isolating indium, gold and rare metals and metal-free glass. Unbroken LCDs were furnace-heated in the absence of oxygen. The exhaust gases were heated at elevated temperatures in order to decompose toxic substances and minimise dioxin formation. The tar was re-used as fuel. The decomposed glass was treated with chemicals to release metals (indium, gold and rare metals). Both glass and metals could then be re-used, not necessarily for the electronics industry.

Nishama Stainless Chemical Company has disclosed a method for regeneration of glass substrates from colour-LCDs by selectively removing the various layers that are attached to the inner surface of a coloured glass filter. A coloured glass substrate usually comprises a black-matrix film, a colour filter, an overcoat film, an indium-tin oxide (ITO) film and a polyimide film, successively formed in this order. The company disclosed a variety of chemical-based peeling solutions that allow sequential removal of adhered layers.[34]

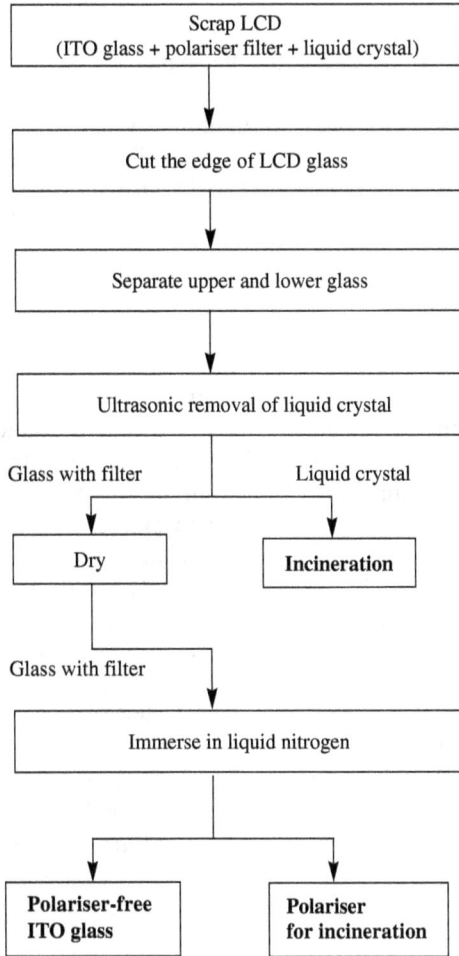

Figure 19 Li and Yang's method for panel disassembly.

A major problem in recycling panels is how best to remove the polarising films. Li and Yang[35] disclosed a method for the recycling of scrap liquid-crystal displays with efficient polariser removal, as shown in Figure 19.

An LCD panel is cut open and the liquid crystal is removed by ultrasound-assisted solvent washing to give glass substrate free from liquid crystal. The latter is incinerated whilst the former is immersed in liquid nitrogen, which releases the polarising films. This method is very expensive and perhaps unfeasible on an industrial scale.

Sharp plc has developed a technique for recovering the LCD glass through crushing, then mixing with clay and feldspar before moulding the slurry into tiles and firing in a kiln. It has also developed techniques for recovery of the plastic components from LCDs.[26]

7 Future Outlook

As LCDs continue to penetrate the flat-panel display market, with increasing year-on-year sales of LCD TVs forecast for the next decade, LCD manufacturing must adopt strategies to minimise waste disposal and maximise recovery and re-use. Future volumes of waste will be significant and the current practice of incineration and/or landfill of LCD panels may need to be reconsidered, as this not only leads to loss of liquid crystal and glass but also indium metal.

7.1 LCD Panels

Indium is a silvery-white rare metal with an estimated abundance of 240 parts per billion by weight in the earth's crust. Indium is found mainly as a by-product in sphalerite, the mineral ore from which zinc is extracted. Even though sphalerite contains indium in the range one to one hundred parts per million only, it is still economically more viable to extract indium from zinc rather than from any other metal ore. Over 55% of the total indium extracted is used by the LCD industry in the manufacture of transparent indium-tin oxide (ITO) electrical conductors. However, mining for indium alone is economically unviable and new sources of indium are dependent on demand for zinc, which unfortunately is not strong at present. The decline in zinc mining, coupled with dramatic growth of the LCD industry, has seen the price of indium per kilogram rocket from a meagre $97 (2002) to a high of $960 (February 2006), with its current price hovering at around $750 (March 2008). The estimated annual amount of indium used globally in the LCD industry is 20 tonnes and, based on the current price of indium, equates to a market value of $15 million.

The cost of the liquid-crystal mixture used in an LCD device will be dependent on its application. The cost of a fast-switching ($<10\,$ms) LC mixture will obviously be more than that of a moderate-switching device. The LCD industry is very competitive, with narrow profit margins. The true cost of LC mixture is a well-kept secret but probably ranges from US$5–15 per gram for TFT applications, with the lower figure being used for LCD computer monitors and the upper figure for high-specification LCD TV applications. However, in today's competitive manufacturing market, it is reasonable to speculate that LC costs in excess of $15 per gram are highly unlikely. Nevertheless, only a few years ago and before the LCD TV boom, the price of LCs for such applications may well have been in excess of $15 per gram, making it comparable with the price of gold at that time (gold price in 2005, $15 per gram). Table 3 shows the value of the LC waste-stream based on different prices.

A UK Government-funded project under the acronym REFLATED is investigating solutions to recovery of polariser, glass, metal and liquid crystal associated with an LCD panel.[36]

Table 3 Estimated values of liquid-crystal mixtures.

LC price (US$) per			Price (US$) per gram	LC value (US$ millions)[a]
cm^2	$inch^2$	m^2		
0.0043	0.027	43	8.5	1032
0.007	0.046	71.5	14.3	1716
0.01	0.065	100	20	2400
0.014	0.092	143	28.6	3432

[a]Based on 48 million m^2 of glass.

7.2 Smart Disassembly

The construction of an LCD device is such that it is not designed for easy disassembly and re-use. Disassembly has to be undertaken manually, which poses contentious health issues with regard to mercury-containing backlights. Breakage of a backlight during manual disassembly will expose the handler to toxic mercury vapour. A possible solution may be to use active-disassembly, using smart polymers.[37] In Japan, smart polymers (temperature- and moisture-sensitive) for active disassembly have been developed to ensure easy separation of glass substrates.[38]

7.3 Legislation

The manufacturers and recyclers need to work closer. At present the LCD-containing WEEE is diverse and is mixed together. LCD devices operate using different technologies and the liquid-crystal mixture employed is technology-specific. In order to ensure efficient recycling of LCD panels, efficient pre-sorting strategies are needed. The industry needs to consider improved labelling of panels to better inform the recycler so as to warrant its best course of action. This may require government intervention through imposed legislation.

References

1. M. Lo, Marketing Director, DisplaySearch Shanghai, Shanghai, China 200051, Data from Display Search Q1'08 Worldwide FPD Forecast Report, personal communication.
2. Sharp Corporation press release, 26/02/08, available at http://www.sharp-world.com/corporate/news/080226.html, accessed 18/04/08.
3. Basel Convention on the Control of Transboundary Movements of Hazardous Wastes and their Disposal, http://www.basel.int/text/con-e-rev.pdf, accessed 18/04/08.
4. Directive 2002/96/EC of the European Parliament and of the Council of 27 January 2003 on waste electrical and electronic equipment (WEEE) – Joint

declaration of the European Parliament, the Council and the Commission relating to Article 9, http://ec.europa.eu/environment/waste/weee/index_ en.tm, accessed 18/04/08.

5. Directive 2002/95/EC of the European Parliament and of the Council of 27 January 2003 on the restriction of the use of certain hazardous substances in electrical and electronic equipment, http://ec.europa.eu/environment/ waste/weee/index_en.htm, accessed 18/04/08.

6. Regulation (EC) No. 1907/2006 of the European Parliament and of the Council of 18 December 2006 concerning the Registration, Evaluation, Authorisation and Restriction of Chemicals (REACH), establishing a European Chemicals Agency, amending Directive 1999/45/EC and repealing Council Regulation (EEC) No. 793/93 and Commission Reg- ulation (EC) No. 1488/94 as well as Council Directive 76/769/EEC and Commission Directives 91/155/EEC, 93/67/EEC, 93/105/EC and 2000/21/ EC, http://ec.europa.eu/environment/chemicals/reach/reach_intro.htm, accessed 19/04/08.

7. W. Becker, Liquid Crystal Newsletter, No. 19, SID 2004 Special, Merck KgaA, Darmstadt.

8. L. Chang, GHGs Inventory and Reduction: Experience of TFT-LCD Panel Manufacturer, AU Optronics Corporation, Corp. ESH Dept, November 4, 2005.

9. F. Reinitzer, *Monash. Chem.*, 1888, **9**, 421; F. Reinitzer, *Liq. Cryst.*, 1989, **5**(1), 7 (English translation).

10. O. Lehmann, *Z. Physik. Chem.*, 1889, **18**, 273.

11. D. Vorländer, *Kristallinisch-flüssige Substanzen*, Enke Stuttgart, 1908.

12. D. Vorländer, *Ber. Dtsch. Chem. Ges*, 1908, **41**, 2033.

13. G. Heilmeier, L. Zanoni and L. Barton, *Appl. Phys. Lett.*, 1968, **13**, 46.

14. M. Schadt and W. Helfrich, *Appl. Phys. Lett.*, 1971, **18**, 127.

15. G. W. Gray, K. J. Harrison and J. A. Nash, *Electron. Lett.*, 1973, **9**, 130.

16. S. J. Woltman, G. D. Jay and G. P. Crawford (ed.), *Liquid Crystals: Frontiers in Biomedical Applications,* World Scientific Publishing Co. Pte. Ltd, 2007.

17. J. B. Park, H. Y. Kim, Y. H. Jeong, United States Patent Application, US 20060279684, 2006.

18. P. Kirsch, *Modern Fluoro-organic Chemistry: Synthesis, Reactivity and Applications,* Wiley-VCH, Weinheim, 2004.

19. M. Hird, *Chem. Soc. Rev.*, 2007, **36**, 2070.

20. G. Friedel, *Ann. Physik.*, 1922, **18**, 273.

21. N. A. Clark and S. T. Lagerwall, *Appl. Phys. Lett.*, 1980, **36**, 899.

22. P. J. Collings and M. Hird, *Introduction to Liquid Crystals,* Taylor and Francis, New York, 1997.

23. G. W. Gray (ed.), *Thermotropic Liquid Crystals (Critical Reports on Applied Chemistry),* John Wiley and Sons, Chichester, vol. 22, 1987.

24. M. L. Socolof, J. G. Overly, L. E. Kincaid and J. R. Geibig, Desktop Computer Displays. A Life Cycle Assessment, vol. 1, December 2001,

EPA-744-R-01-004a; http://www.epa.gov/dfe/pubs/comp-dic/lca/index.htm, accessed 18/04/08.

25. E-Waste Curriculum Development Project, Phase 1: Research and Content Development – Literature Review, The Natural Edge Project, July 2006; http://www.naturaledgeproject.net/Documents/E-Waste%20Literature% 20Review%20-%20FINAL.pdf, accessed 18/04/08.

26. 2008 Review of Directive 2002/96 on Waste Electronic and Electrical Equipment (WEEE): Final Report, United Nations University, Germany, August 2007; http://ec.europa.eu/environment/waste/weee/pdf/final_rep_unu.pdf, accessed 18/04/08.

27. Literature Review, Flat Panel Displays: End of Life Management Report, Final Report, King County Solid Waste Division, Seattle, USA, August 2007.

28. P. Bocko, Corning Investor Meeting at Taichung, 2006.

29. D. Hsieh, Display Trends: Display Devices, Spring 2006; http://www.displaysearch.com/cps/rde/xbcr/SID-0A424DE8-644765D7/displaysearch/DD_spring06.pdf, accessed 18/04/08.

30. M. Nakamichi, Jpn. Kokai. Tokkyo Koho JP 2005 227,508 (Cl. GO2F1/13), 25 Aug 2005, Appl. 2004/35,597, 12 Feb 2004. Decomposition method and apparatus for liquid crystals in recycling of liquid crystal panels, CAN: 143: 219606g.

31. H. Hiroshi, T. Kiyobumi and T. Haruyoshi, Dainippon Ink & Chemicals, Publication number JP2006091266, Application number JP20040274908 20040922, 2006.

32. H. Hiroshi, T. Kiyobumi and T. Haruyoshi, Dainippon Ink & Chemicals, Publication number JP2006089519, Application number JP20040273184 20040921, 2006.

33. I. Sumoge and F. Hiruma, Densho Engineering Co. Ltd, Saitama, Japan, *Denki Kagakkai Gijutsu, Kyoiku Kenkyu Ronbunshi*, 2005, **12**(2), 33, CAN 146:445162.

34. T. Nishiyama, M. Deguchi, K. Itokawa, K. Mizoguchi, Nishama Stainless Chemical Co. Ltd, Japan, Jpn. Kokai Tokkyo Koho (2003), Publication number JP2003279915; Application number JP2002377810 20021226, CAN 139:268132.

35. C.-H. Li and K.-L. Yang, Publication number TW527232B, Application number TW20010200570 20010110, 2003.

36. DIUS Report, Innovation Nation, 2008, http://www.dius.gov.uk/publications/ScienceInnovation.pdf, accessed 18/04/08.

37. J. D. Chiodo, J. McLaren, E. H. Billett and D. J. Harrison, *IEEE*, 2000, **318**.

38. DTI Global Watch Mission Report, Waste Electrical and Electronic Equipment (WEEE): Innovating novel recovery and recycling technologies in Japan, 2005.

The Role of Collective versus Individual Producer Responsibility in e-Waste Management: Key Learnings from Around the World

MARK DEMPSEY AND KIRSTIE MCINTYRE

1 Introduction

In countries where e-waste management has been regulated, various practical policy models have been adopted. This chapter examines those different models and describes their advantages and disadvantages from an extended-producer-responsibility perspective. In particular, collective-producer responsibility and individual-producer responsibility are compared and evaluated. The chapter also reviews competitive *versus* non-competitive systems of compliance.

Section 1 describes why e-waste is an issue and why it requires attention from legislators, manufacturers and consumers, and provides a background to Producer Responsibility legislation.

1.1 E-waste and Its Environmental Impacts

E-waste legislation stems from growing concerns regarding the environmental impacts of this waste stream. Electrical equipment contains materials that can cause environmental problems if they are disposed of to landfill or incinerated. Hazardous substances are contained within components such as printed-circuit boards, cables, wiring, plastic casing containing flame retardants, display equipment, including cathode-ray tubes, batteries and accumulators, capacitors, resistors and relays, and connectors. The landfilling of WEEE risks the

Issues in Environmental Science and Technology, 27
Electronic Waste Management
Edited by R.E. Hester and R.M. Harrison
© Royal Society of Chemistry 2009
Published by the Royal Society of Chemistry, www.rsc.org

leaching of heavy metals, including lead, cadmium and mercury, into groundwater or the evaporation of mercury into the air.

These concerns are growing as the WEEE waste stream shows growth in Europe of between 3 and 5% per annum. The latest figures state that 10.3 million tonnes/year (equivalent of approximately 19 kg/inhabitant) of Electrical and Electronic Equipment (EEE) is put on the European market.[1] In addition to this, 8.3–9.1 million tonnes/year of Waste Electrical and Electronic Equipment (WEEE or e-waste) results. However, only 2.2 million tonnes was estimated as being collected and treated in 2005 (approximately 5 kg/inhabitant) within the EU, amounting to only a 25% recovery rate of WEEE.[1] Pichtel[2] states that 'European studies estimate the volume of e-waste is rising by 3–5% per year, almost three times faster than the Municipal Solid Waste stream'. There are a number of factors driving these figures:

- The increased uptake of technology and increased expenditure on IT equipment which has penetrated the whole globe
- The numbers of PCs globally are growing at an exponential rate (see Figure 1)
- Economies of scale in production have resulted in equipment being more affordable and accessible
- The reduced in-use life of IT equipment, and rapid developments and advancements to IT products, have resulted in consumption keeping pace with technology.

This, combined with price reductions in IT equipment, means that upgrades are both affordable and in some cases are included in promotions (*e.g.* with cellular-phone contracts). This results in increased numbers of pieces of IT equipment becoming obsolete, and at a faster pace. Pichtel[2] predicts that 'the average lifespan of a Pentium-class computer is currently 2 to 3 years and will gradually decline'.

1.2   Background to Producer Responsibility

Producer Responsibility is part of a new generation of environmental policy instruments aimed at governing the disposal of products when they come to the end of their usable lives. These new environmental laws are having, and will increasingly have over the coming years, a substantial impact on governments, manufacturing industry, local authorities and waste management systems across the globe.

The producer responsibility concept was put forward and discussed by academics in the early 1990s. From the outset, producer responsibility was considered as a means of creating design incentives for manufacturers, as well as requiring producers to take responsibility for the end-of-life costs of their products. The extent to which legislation has provided incentives for producers to improve the design of their products will be explored later in this chapter.

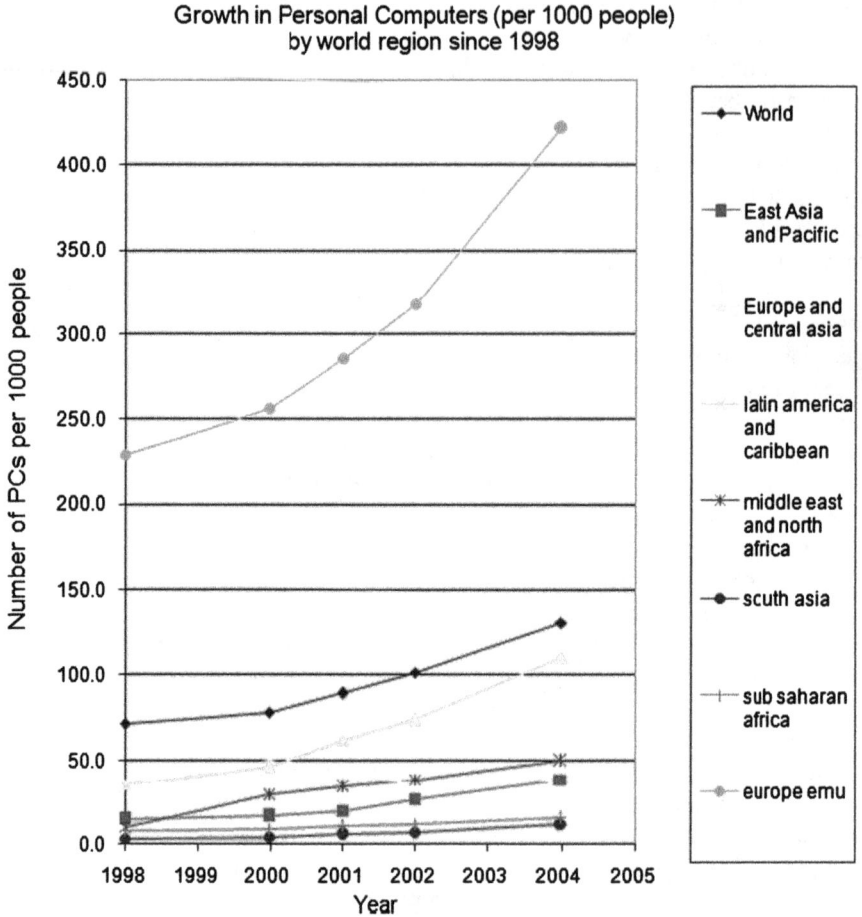

Figure 1   Growth in personal computers per 1000 people since 1998. (Source: adapted from figures in WDI, 2000,[17] 2006,[18] and Little Data Book 2003,[19] 2004,[20] 2005[21].)

Since the first Producer Responsibility systems were developed in Scandinavia and Germany in the early 1990s, other countries across the globe have adopted this approach to tackle environmental problems within their own countries. Producer Responsibility systems are now a key part of the waste policy toolkit and systems are in place or in development across Europe, Asia and North America.

The implementation of Producer Responsibility illustrates that environmental benefits and economic costs vary widely, dependent upon the approaches adopted by individual nation states. Learning lessons from past implementation is crucial to the development of Producer Responsibility systems, which achieve the objectives of sustainable development whilst enabling competitive and low-cost compliance.

One of the common waste streams that has been a focus for Producer Responsibility legislation is waste electronic and electrical equipment (WEEE). This chapter contrasts the different approaches taken by governments to implement Producer Responsibility for waste electronic and electrical equipment (WEEE).

1.3   Defining Individual and Collective Producer Responsibility

Under Collective Producer Responsibility (CPR), all manufacturers are collectively responsible for e-waste arising from all products. Under CPR, responsibility is allocated *via* the market share of producers. Under Individual Producer Responsibility (IPR), each manufacturer is responsible for the waste arising from its own products.

Individual Producer Responsibility (IPR) was designed to provide incentives to producers for taking responsibility for the entire lifecycle of his/her own products, including end of life.

IPR can be achieved in a number of ways, including return-share systems, brand segregation or through own-brand recycling systems. These are reviewed in greater detail below.

IPR does not require each producer to have a separate infrastructure for the collection and treatment of only their own-brand appliances. Operational recycling systems can be arranged to encompass individual financial producer responsibility.

The distinction between IPR and CPR and individual and collective recycling systems is illustrated by Figure 2. This provides a model for the classification of Member States' producer responsibility systems in the following sections.

The variation in the type and amount of individual control experienced by a producer is usually described in terms of financial and physical responsibility. Rossem, Tojo and Lindhqvist[3] provide definition to these two forms of IPR:

A producer bears an individual financial responsibility when he/she pays for the end-of-life management of his/her own products. A producer bears an individual physical responsibility when 1) the distinction of the products are made at minimum by brand and 2) the producer has control over the fate of their discarded products with some degree of involvement in the organisation of the downstream operation.

Individual financial producer responsibility can be delivered by a collective recycling system (B1 and B2). This approach requires the cost to the producer of the recycling of its WEEE to be differentiated to reflect the relative cost of end-of-life management. For example, sampling of WEEE is used to estimate the quantities of a specific producer's products that flow through a collection and processing system.

When individual producers operate their own-brand recycling systems (B3), brands are physically separated and recycled by their original producer. If the

<table>
<tr><th rowspan="2">Financial responsibility</th><th colspan="2">A: Collective Producer Responsibility:</th><th colspan="2">B: Individual Producer Responsibility:</th></tr>
<tr><td colspan="2">Based on market share of producers.

E.g. legal requirement within WEEE Directive for historic waste.</td><td colspan="2">Based on amount of producer's products returned for recycling.

E.g. legal requirement within WEEE Directive for future waste; legal requirement within Maine and Washington legislation.</td></tr>
<tr><th rowspan="4">Recycling system</th><td rowspan="2">A1: Non competitive

Monopolistic schemes

E.g. Sweden, the Netherlands, Belgium, Switzerland</td><td colspan="2">Competitive:

E.g. Germany, the UK, France</td><td>B1: Non competitive

Monopolistic schemes

E.g. ICT Milieu 1999–2002</td><td colspan="2">Competitive:

E.g. Japan</td></tr>
<tr><td>A2: Collective recycling system</td><td>A3: Individual recycling system</td><td rowspan="2"></td><td>B2: Collective recycling system</td><td>B3: Individual recycling system</td></tr>
<tr><td>*E.g.* Compliance schemes such as ERP</td><td>*E.g.* Direct take-back events or systems</td><td>*E.g.* Group A and B compliance schemes</td><td>*E.g.* PC recycling in Japan</td></tr>
</table>

Figure 2   Model to distinguish between individual and collective producer responsibility and individual and collective recycling systems.

producer is able to design their products to minimise end-of-life costs and environmental impacts, the producer will obtain the financial benefits that accrue from this investment.

2   The WEEE Directive in Europe

This section describes the current legislative and voluntary approaches to addressing e-waste in a selection of regions across the world.

2.1   The WEEE Directive's Approach to Individual and Collective Producer Responsibility

The WEEE Directive[4] introduces producer responsibility for electrical and electronic waste. It establishes mass-based recycling and recovery targets imposed on all EU member countries. European governments are responsible for ensuring a minimum collection target of WEEE, currently set at 4 kg per capita per year, free of charge to consumers. Then, producers are responsible for proper treatment of all collected products so that the imposed recycling and recovery targets are met.

Table 1 WEEE Directive categories and recovery and recycling targets.

Category	Equipment	Recovery target	Recycling target
1.	Large household appliances (*e.g.* washing machines, cookers)	80%	75%
2.	Small household appliances (*e.g.* toasters, irons, hairdryers)	70%	50%
3.	IT and telecommunications equipment (*e.g.* PCs, copiers, phones, mobiles)	75%	65%
4.	Consumer equipment (*e.g.* TVs, videos, hi-fis)	75%	65%
5.	Lighting equipment (*e.g.* fluorescent lamps, but excludes filament light bulbs)	70%	50%
6.	Electrical and electronic tools, exception large-scale stationary industrial tools (*e.g.* lawnmowers, sewing machines, drills)	70%	50%
7.	Toys, leisure and sports equipment (*e.g.* video games, bike computers, slot machines)	70%	50%
8.	Medical devices, except implanted and infected products (*e.g.* ventilators, analysers)	To be decided in 2008	To be decided in 2008
9.	Monitoring and control instruments (*e.g.* smoke detectors, thermostats)	70%	50%
10.	Automatic dispensers (*e.g.* drinks dispensers, chocolate dispensers, ATMs)	80%	75%

The WEEE Directive applies to importers, producers, retailers and users of EEE, and to businesses that treat or recover WEEE. The WEEE Directive applies to electrical and electronic equipment (EEE) in ten product categories listed in Table 1.

This legislative structure has been designed considering two basic assumptions: (i) by making producers responsible, an economic incentive is created to improve the environmental design of products and (ii) recycling and recovery of large quantities of waste should lead to a reduction in environmental impact.[5]

The original assumption of the European Commission in implementing WEEE has been that: 'This financial or physical responsibility creates an economic incentive for producers to adapt the design of their products to the prerequisites of sound waste management.'[6] This objective is further solidified within the preamble of the WEEE Directive. The WEEE Directive[4] states that:

'The establishment, by this Directive, of producer responsibility is one of the means of encouraging the design and production of electrical and electronic

equipment which take into full account and facilitate their repair, possible upgrading, reuse, disassembly, and recycling'

2002/96/EC: Recital 12

'In order to give maximum effect to the concept of producer responsibility, each producer should be responsible for financing the management of the waste from his own products'.

2002/96/EC: Recital 20

Article 8 of the WEEE directive distinguishes between historical[i] and future[ii] waste and the way these streams are treated. With respect to historical waste, all producers existing in the market contribute proportionally to the management of WEEE – according to their *current market share* by type of equipment. For future waste, on the other hand, each producer is responsible for the WEEE from his/her own products. Article 8.2 states:

For products put on the market later than 13 August 2005, each producer shall be responsible for financing the operations referred to in paragraph 1 relating to the waste from his own products. The producer can choose to fulfil this obligation either individually or by joining a collective scheme.

Therefore Article 8.2 establishes IPR for the recycling of products put on the market after 13 August 2005. Making each producer responsible for financing the end-of-life costs of their own-branded products enables end-of-life costs to be fed back to the individual producer. By modifications to the product design, the producer can directly influence the end-of-life cost. Without IPR these incentives for design improvements are lost.

2.2   Implementation of Individual and Collective Producer Responsibility in the EU

The WEEE directive became European law in 2003 and has been implemented by the European Union Member States during the period 2005–2007. The WEEE Directive has faced important implementation problems. The Directive is implemented in 27 countries and each country has its own system. This stems from the fact that the EU can impose Directives on the member countries, but the latter have certain freedoms as to how they translate Directives into national laws. National (or even regional) laws can be stricter, and implementation details may differ substantially. As a consequence, companies

---

[i] Historical WEEE refers to EEE products that have been placed on the market prior to the WEEE Directive coming into force – set as 13 August 2005 – and subsequently become waste.
[ii] Future, or sometimes referred to as New, WEEE means EEE products that are placed on the market on or after 13 August 2005 and subsequently become waste.

struggle with inter-country differences and unclear specifications on how to comply.

Collective Producer Responsibility for WEEE has been implemented in all 27 Member States of the European Union. Manufacturers (or producers) declare their current sales figures to the authorities. These are totalled and divided to give a percentage market share, *e.g.* for producer $X = 20\%$. This means that producer X is responsible for 20% of the WEEE returning. This means that producer X pays for the collection, treatment and recycling of 20% of any manufacturer's equipment, not necessarily their own or even in their own product category. The operational systems have taken the form of either non-competitive (A1) or competitive recycling systems.

Many countries have failed to transpose the IPR concept into their national laws. According to an assessment made by a coalition of NGO and industry actors,[7] only 13 out of 27 countries have transposed this article adequately while another 4 have transposed it only partially (see Table 2). Ten Member States (MS) have not transposed 8(2) as intended in the WEEE Directive. Instead, the legislation in these countries makes producers jointly responsible for the recycling of future products, making it impossible to implement IPR.

Similarly, a more recent study undertaken by Okopol[8] for the European Commission came to comparable conclusions regarding the outcome of transposition on Article 8.2. In this study, 9 of the 27 MS have, in the contractor's opinion, correctly transposed the requirements of IPR for future WEEE. Eight out of 27 MS have transposed Article 8(2) in an ambiguous manner, leaving considerable doubt as to whether IPR is legally binding in the national legal text. Likewise, the report finds that 10 out of 27 MS have simply missed the requirement for each producer to finance the waste from his/her own products.

2.3 ICT Milieu, The Netherlands

It is interesting to examine the only example of IPR to have existed within Europe. ICT Milieu is one of the two collection systems for WEEE in the Netherlands. ICT Milieu collects IT equipment, including printers and

Table 2 Analysis of the transposition of IPR in the EU.

Member States which have transposed Article 8.2	Belgium, Cyprus, Czech Republic, Estonia, Ireland, Italy, Lithuania, Luxembourg, Malta, Romania, Slovakia, Sweden and the Netherlands
Member States which have inadequately transposed Article 8.2	Bulgaria, Denmark, Finland, France, Greece, Latvia, Portugal, Slovenia, Spain, and the UK
Member States which have partially transposed Article 8.2	Austria, Germany, Hungary, Poland

Source: Iprworks.org, 2007[8]

telecommunication equipment. White and brown goods are collected and managed by NVMP, the other organisation in The Netherlands managing WEEE.

In the Dutch system, before the WEEE Directive was enacted, up until 1 January 2003, individual producers received a monthly invoice directly from the recycler based on the weight of the recycled products. Every container delivered to the recycling company was hand-sorted. Each product was weighed on a scale. Brands were visually identified and each unit was assigned to a manufacturer and logged using a touch-screen panel. Any name which was not part of ICT Milieu was registered as a 'free rider'. The cost of manual sorting was a few cents/kg. Making it weight-based made the calculation of the recovery cost simple and clear. Payment for returned products was made at the moment product recycling took place.

ICT Milieu therefore provided an example of IPR achieved through a non-competitive 'collective' recycling system (B1).

3   E-waste Laws and Voluntary Agreements in Other Countries

Section 3 reviews e-waste systems across the world. The European Union may be leading the way with its WEEE Directive, but many other countries have implemented or plan to implement their own e-waste laws or voluntary requirements. Hewlett-Packard estimates that, by 2010, 75% of its global sales will be in countries with e-waste legislation. The spread of this form of law can be seen in Figure 3.

3.1   Japanese Electronics Take-back Directive

The Japanese directive came into force in April 2001. This directive sets treatment standards *via* a waste management law. The directive's scope is

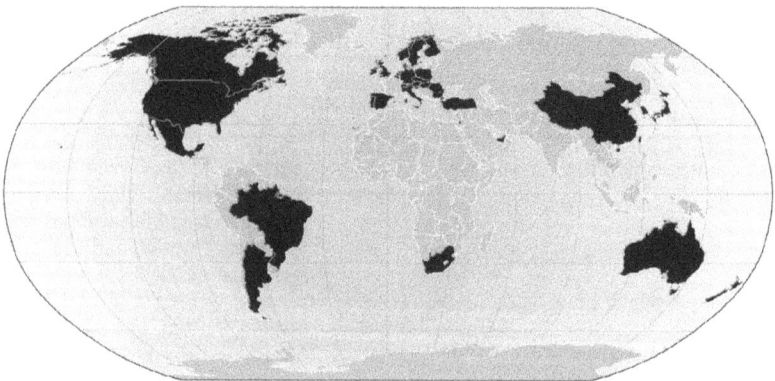

Figure 3   Countries with e-waste legislation or voluntary agreements. (Source: McIntyre, 2007[22].)

limited to TV sets, cooling devices, washing machines and air conditioners. Treatment of PCs is covered in a different law.

This directive ensures that producers pay for the end-of-life management of their own products through a return-share system. Retailers collect the returns. Producers are allowed to charge an end-of-life management fee to the end-user. This fee is collected by the retailers and managed by individual manufacturers through the management of a common recycling ticket. From the money collected, the producers pay the recycling plants, depending on how many products have been recycled.

In contrast to Europe, in Japan the producers have the operational responsibility for recycling facilities. Recycling plants are owned by two collective systems, systems A and B. Collected returns are separated according to brands and sent to recycling centres corresponding to the collective system A or B. Manufacturers' processing fees may differ, depending on product specifications, and environmental savings remain with the manufacturers. This creates incentives for the manufacturers to design recyclable products. The incentive is driven by the competition between the two collective systems. If manufacturers in one collective system could recycle their products more cheaply, they would get a cost advantage over the competitors in the other collective system.

There is a separate producer responsibility system governing the recycling of PCs. The PC recycling system operates through the nationwide postal service. Consumers send their used products *via* post to producers' own recycling plants. In this way an own-brand IPR system has been established.

Therefore, according to our model, Japanese producer responsibility can be classified as B2 for TV sets, cooling devices, washing machines and air conditioners, and B3 for PCs.

3.2   Product Take-back in the USA

The producer responsibility legislation in Maine is limited to household computers and video displays over 4-inch size, CRT display monitors, TV sets, laptop computers and portable DVD players. Producer responsibility legislation has been enacted as of January 1 2006.

The Maine directive[9] states that:

*Each computer monitor manufacturer and each television manufacturer is **individually responsible** for handling and recycling all computer monitors and televisions that are produced by that manufacturer or by any business for which the manufacturer has assumed legal responsibility, that are generated as waste by households in this State and that are received at consolidation facilities in this State. In addition, each computer manufacturer is responsible for a pro rata share of orphan waste computer monitors and each television manufacturer is responsible for a pro rata share of orphan waste televisions generated as waste by households in this State and received at consolidation facilities in this State.*

In this system the collection of waste is handled by municipalities. The waste is then passed to a consolidator. The number of consolidators in Maine is currently seven. Manufacturers of e-waste are then required to submit a collection/recycling plan that defines the way they satisfy the requirements posed by the legislation. Producer responsibility is based on the number of its products collected for recycling. This is known as return share. In Maine this is determined by a full count of the products collected by consolidators. The manufacturers can choose to (i) recycle a return share of mixed branded e-waste or (ii) pay the consolidator to undertake the recycling of the producers return share, or (iii) recycle the producers' own-branded products. In each of these scenarios the producer must take responsibility for a proportion of orphan products.

The Washington directive is similar to the Maine directive in many aspects. All monitors, TVs or other video displays over 4 inches, desktop computers and laptop computers are covered by this directive. The scope covers products from covered entities only: households, small businesses, charities, local governments and school districts. Producer responsibility legislation will be active as of January 1 2009.

Manufacturers must register with the Department of Ecology. All manufacturers must participate in an approved recycling plan. Manufacturers may join the Standard Plan to manage and finance recycling programs or they may start an independent plan on their own or with others (if the combined return share is above 5% of e-waste.)

The Washington directive defines return share as a manufacturer's percentage, by weight, of identified brands of covered electronic products returned for recycling. The law directs the Department of Ecology (DE) to determine the return share for each manufacturer participating in either the Standard Plan or an Independent Plan. In the first year the Department has sourced return-share data from the Brand Data Management System developed by the National Centre for Electronics Recycling (NCER). In future years the return-share will be calculated statistically by sampling, unlike Maine, where return-share is calculated by a full brand count.

According to our model, both Maine and Washington are classified as B2 systems. However, both States enable B3 systems to exist in parallel. There is a further refinement in that Washington achieves return share through sampling of the waste stream whereas Maine performs a full brand count.

3.3   Product Stewardship in Australia

In Australia the term Product Stewardship has been adopted in preference to Producer Responsibility to describe initiatives to encourage producers to take responsibility for the post-consumer stage of their products. Policy has been primarily driven by the Environmental Protection and Heritage Council (EPHC), which ensures that the federal government, states and territories work together to develop measures to protect the environment. In December 2004, EPHC launched an *Industry Discussion Paper on Co-Regulatory Frameworks for Product Stewardship*,[10] and has since co-ordinated Product Stewardship initiatives for packaging, tyres, televisions and computers.

The Australian governments have identified electronics as a primary waste stream for Product Stewardship. The scope of products under attention is narrower than the WEEE Directive, focusing on IT and TVs. Televisions have been a focus for Product Stewardship as a result of the hazardous nature of waste televisions and the resultant damage to the environment caused by their disposal. As a result the Consumer Electronics Suppliers Association (CESA), in partnership with the Australian Electrical and Electronic Manufacturers Associations (AEEMA) and Product Ecology Ltd, commenced work on developing a proposal for Product Stewardship to cover the televisions waste stream.

A Product Stewardship Agreement is currently being developed in negotiation with the EPHC, and is contingent upon co-regulatory action to develop and enforce legislation to regulate free-riders. The agreement will ensure all States and Territories are covered by the end of a five-year period. The agreement also sets separate collection and recycling targets, which will result in 80% collection and 85% recycling of the material collected by the end of the five-year period of operation.

CESA has also established a Producer Responsibility Organisation entitled *Product Stewardship Australia* (PSA). The agreement proposes that PSA will develop, co-ordinate and promote a collection and recycling scheme for televisions. PSA has been established as a not-for-profit company and will have sole responsibility for organising the recycling of televisions on behalf of producers, collecting and publishing data and developing a public-awareness-raising campaign. PSA will fund its activities through an environmental levy collected from participating television companies, based on market share (model A1).

In 2004 the Australian Government proposed a *National Electrical and Electronic Products Recovery Program*, including a product levy on the sale of new computers, peripherals and TVs, managed by one or more Producer Responsibility Organisations. The computer sector, through the Australian Information Industry Association (AIIA), has been working for several years to develop a proposal for Product Stewardship for end-of-life computers. However, producers have expressed concern about the financial impact of taking responsibility for the entire IT waste stream, including orphan and historical products. The large amount of unbranded equipment supplied by white-box producers results in a significant amount of orphan product at the point of disposal. Ensuring industry equity in cost-sharing in such a diverse sector has proved to be a major stumbling block in negotiations.

4 Discussion

4.1 Competition in E-Waste Management

The way in which Member States implement e-waste recycling systems can have significant impacts on their environmental and economic performance. As illustrated in Figure 2, the compliance approaches for WEEE from private households in Member States can be categorised into two systems: non-competitive and competitive.

Several countries in Europe have implemented a single national-compliance system for WEEE compliance (model A1). Single National-Compliance systems have been the standard approach for countries with legislation prior to the implementation of the WEEE Directive. Producers in Sweden, Norway, Belgium, the Netherlands and Switzerland are required to join Producer Responsibility Organisations (PROs).

While these schemes are very adept at collecting and treating WEEE, concern has been expressed that they are monopolies and that an in-built lack of competition has led to higher recycling costs. Although these schemes can tender for recycling and transport services, producers are faced with no choice as to who they join to demonstrate their compliance.

Lowest cost is important to manufacturers who wish to compete at end of life. Hewlett-Packard is among many companies who wish to compete throughout the product's lifecycle. Manufacturers compete on raw-material costs, manufacturing and operational costs, distribution, service and, of course, end of life. The imposition of a legal framework at end of life does not preclude continuing to compete at this point too.

The European Recycling Platform (ERP) was established by Hewlett Packard, P&G, Electrolux and Sony Europe in order to encourage the creation of competition within Member States compliance systems (model A2). In Germany, France, Spain, the UK and several other countries, competitive compliance schemes have been established (see Table 3).

In order to facilitate competition between different compliance schemes some countries have established a clearing house. This is a body that allocates

Table 3 Countries adopting a single compliance system and competitive compliance systems.

Single compliance system (model A1)	Competitive compliance system (model A2)
Sweden	Denmark
Belgium	Ireland
Luxembourg	Italy
Greece	France
Estonia	Austria
Malta	Germany
Cyprus	Finland
Netherlands	Portugal
Norway	Spain
Switzerland	Slovenia
	Slovakia
	Poland
	Lithuania
	Latvia
	Hungary
	UK

waste arising at municipal sites to compliance schemes based on the obligation of their member companies. This ensures that all separately collected WEEE is recycled, and ensures that producers take responsibility for a fair allocation of different types of collection site. Other countries have facilitated competition by a simplified clearing house or through other means such as tradable evidence notes.

In 2005 Veit[11] examined the different pricing strategies that are seen across the various implementation regimes in Europe. Using a basket of products the average cost of recycling in non-competitive systems is €6.33 per product WEEE, more than double the €2.93 per product average cost of competitive schemes.

In 2006 HP published data underlining that where there is a more-competitive environment for electronics recycling providers, the take-back and recycling costs of retired electronic equipment are lower. Table 4 illustrates that prices are significantly lower in Spain, Austria and Germany, which have competitive recycling systems, compared to costs in the Netherlands, Belgium, Norway and Switzerland, which have non-competitive compliance systems. For example, to recycle a PDA (Personal Digital Assistant) costs consumers less than 1 euro-cent in Spain, 2 euro-cents in Austria, 41 euro-cents in Belgium and 1.33 euros in Switzerland.

4.2   Collective Producer Responsibility: Benefits and Disadvantages

Even though monopolistic collective systems can be more expensive ways in which to comply with e-waste legislation, collective systems themselves are not necessarily bad for the environment, consumers or manufacturers. Indeed many collective systems are achieving high collection and recycling rates, well in excess of the WEEE Directive targets of 4 kilograms per inhabitant.

However, with CPR there is no differentiation of the recycling costs according to how easy the product is to recycle. The costs are based upon the market share of the producer. Therefore, the costs of recycling will be the same for a product that has been designed to be easier to recycle, and a product that is much more difficult to disassemble and recycle.

If recycling costs are financed collectively (*e.g.* according to market share), manufacturers are more likely to focus only on, and minimise, the production costs. If recycling costs are increased due to a particular design modification these costs are absorbed jointly by all producers. Therefore, collective responsibility (models A1, A2 and A3) does not provide an incentive to a producer to design products to be easier to recycle.

4.3   Individual Producer Responsibility: Benefits and Disadvantages

IPR is a policy tool that provides incentives to producers for taking responsibility of the entire lifecycle of his/her own products, including end of life.

Mark Dempsey and Kirstie McIntyre

Table 4 Comparison of recycling costs in 2006 between monopolistic and competitive take-back structures.

Equipment type	Hand-held	Digital camera	Laptop computer	Desktop computer	Consumer inkjet printer	Laserjet printer	Flat screen monitor
Take-back fee in Belgium[a]	0.41 €	1.24 €	1.65 €	2.48 €	1.65 €	1.65 €	4.96 €
Take-back fee in Switzerland[b]	1.33 €	1.00 €	6.00 €	6.00 €	3.00 €	4.00 €	6.00 €
HP take-back cost Norway[c]	0.05 €	0.10 €	1.52 €	3.80 €	1.33 €	4.56 €	3.42 €
HP take-back cost Sweden[c]	0.03 €	0.06 €	0.88 €	2.20 €	0.77 €	2.64 €	1.98 €
HP take-back cost Netherlands[c]	0.03 €	0.07 €	1.08 €	2.70 €	0.95 €	3.24 €	2.43 €
Cost of take-back in Spain (ERP)[d]	0.01 €	0.01 €	0.20 €	0.50 €	0.18 €	0.75 €	0.81 €
Cost of take-back in Austria (ERP)[e]	0.02 €	0.02 €	0.39 €	0.83 €	0.34 €	1.00 €	1.49 €
Cost of take-back in Germany (ERP)[f]	0.01 €	0.01 €	0.07 €	0.38 €	0.12 €	0.43 €	0.29 €
Average weight/unit sold	0.125 kg	0.250 kg	4 kg	10 kg	3.5 kg	12 kg	9 kg

All figures expressed in euros. 1.5 CHF = 1 euro.
All figures excluding VAT.
[a] Prices charged by RECUPEL.
[b] Take-back and recycling prices set by SWICO (including packing and battery take-back, internal administration and programs for waste reduction). VFG includes cost of additional services (*e.g.* packaging and battery take-back).
[c] Take-back and recycling cost charged by National Consortium to HP, distributed on all HP products sold the same period.
[d] Prices of ERP (ERP reported cost per kilo multiplied by average weight of unit sold). Prices of other national take-back and recycling systems are slightly different.
[e] Prices of ERP (ERP reported cost per kilo multiplied by average weight of unit sold). Prices of other national take-back and recycling systems are slightly different. Include payments to municipalities, communication and 'Waste Minimisation Project'.
[f] Based on charges of ERP (ERP reported cost per kilo multiplied by average weight of unit sold), distributed on all HP products sold the same time. Prices of other take-back and recycling systems in Germany may differ. Cost of national register (EAR) included.
Source: Hewlett Packard, 2006[16]
no g or h notes.

Making each producer responsible for financing the end-of-life costs of their own-branded products enables end-of-life costs to be fed back to the individual producer. IPR provides a competitive incentive for producers to design their products so that they are easier and therefore cheaper to recycle. This was the intention of European policy makers when they developed the WEEE Directive.

Without IPR these incentives for design improvements are lost. Producers are not rewarded for making their producers easier to recycle as the end-of-life costs are related to market share of sales rather than the costs of end-of-life management of producer's products.

IPR creates a strong link between the waste product and the producer. IPR allows the manufacturer to get feedback about the end-of-life issues related to the product. The recycling plants provide the manufacturer with product design-related feedback from the recycling of their own product. Feedback reports from the recyclers encourage proposals for design improvements on issues such as material composition, ease of disassembly and labelling.

The disadvantages of IPR systems are much discussed by those who do not agree that they achieve the environmental objectives of the WEEE directive. However, since many EU member states have only recently completed their national implementations of the Directive, it may be premature to draw such conclusions.

A significant consideration to implementing IPR is the need for financial guarantees. The WEEE Directive refers to these in conjunction with future WEEE. The premise is that without collective responsibility, the burden of collecting, treating and recycling WEEE will fall on society if a producer withdraws from the market. Therefore each producer within an IPR system needs to make a financial guarantee that would survive such an eventuality. The funds or guarantees set aside would be used to meet that producer's continuing collection and recycling obligations after they had exited the market. The specific type of guarantee needed will depend on the specific details of each recycling operation once implemented. To be effective, a financial guarantee must ensure that the costs of collection and treatment of a producer's products falls neither on producers that did not produce them nor on the public.

Producers complying through a collective market-share-based calculation are not required to establish a financial guarantee. This creates a disadvantage for producers wishing to comply through IPR.

4.4 Evaluating Collective **versus** Individual Producer Responsibility

There are a number of factors which make a comparison of the performance of collective and IPR subject to errors. There are fewer examples of IPR systems to evaluate. In Europe IPR has not been implemented, and therefore an evaluation of its effectiveness in the European context is not possible. Some data are available from IPR systems in Japan. However IPR systems in North

America are more recent and therefore data availability for these systems is reduced. Finally the scope of legislation in Japan and North America is different from the scope of the WEEE Directive.

Nevertheless, this section attempts to provide an evaluation of IPR compared to CPR systems. Four performance indicators have been selected, which align with the original objectives and targets of the WEEE Directive.

4.4.1 Collection Rates

Table 5 compares the collection performance between Japan, which operates an IPR system, and European countries, which currently operate different forms of CPR. The data represent collection of products classified in Category 1 of the WEEE Directive (Large Domestic Appliances).

Japan achieved 2.58 kg/inhabitant of Category 1 products despite a narrower scope than the WEEE Directive. This matches or exceeds Austria, Czech Republic, Estonia, Hungary, the Netherlands and Slovakia. Japan achieved 0.82 kg/inhabitant in Category 4 despite narrower scope. This matches or exceeds Austria, Czech Republic, Estonia, Hungary, Ireland, Slovakia and closely matches the EU average (0.88 kg/capita).

Table 5 A comparison of the collection performance between Japan, which operates an IPR system, and European countries.[1]

For Category 1	Collection (kg/capita)	Scope
Japan	2.58	Data for fridges/freezers, air conditioners and washing machines:
Czech Republic	0.14	Large cooling appliances
Slovakia	0.35	Refrigerators
Estonia	0.48	Freezers
Hungary	0.91	Other large appliances used for refrigeration, conservation and storage of food:
Austria	2.00	Washing machines
Netherlands	2.59	Clothes dryers
Belgium	2.99	Dish washing machines
Euro average	3.11	Cooking
Finland	4.75	Electric stoves
Sweden	5.01	Electric hot plates
Ireland	6.68	Microwaves
UK	7.17	Other large appliances used for cooking and other processing of food: Electric heating appliances Electric radiators Other large appliances for heating rooms, beds, seating furniture: Electric fans Air conditioner appliances: Other fanning, exhaust ventilation and conditioning equipment

This illustrates that IPR and CPR achieve comparable levels of collection. In Europe and Japan producers are not directly responsible for the collection of WEEE. Therefore, it is likely that other factors, such as the extent of the collection infrastructure, are the key determinants of collection rate.

4.4.2 Recycling Rates

Table 6 provides a comparison between the recycling performance of the Netherlands and Belgium, which have mature CPR systems (model A1), and Japan, which has established an IPR return-share system for electronic recycling (B2) and an own-brand IPR system for PC recycling (B3).

This illustrates that recycling levels are high in countries with CPR and IPR and have exceeded national targets.

4.4.3 Impact on Product Design

Tojo (2006)[12] analysed the design benefits of the Japanese e-waste recycling system. This showed that in Japan the IPR system has led to the following benefits:

- Use of Design for Environment assessment tools including end-of-life phase
- Marking of materials and locations for ease of dismantling
- Unification of materials (plastics, magnetic alloys)
- Reduction of the number of components and screws
- Standardisation of screws
- Use of recycled plastics in new components
- Development of recycling technologies
- Separation of various types of plastics
- Tools for ease of manual dismantling
- Communication between recyclers and designers

Table 6 A comparison of the recycling performance between Japan, the Netherlands and Belgium.

	Netherlands (2001)	*Belgium (2003)*	*Japan (2006)*
Large domestic appliances	85	84	78
Refrigerators and freezers	74	81	64
TVs	80	83	77
Small domestic appliances and ICT	60	82	
Air conditioners			87

Source: Data for Netherlands and Belgium from Bio Intelligence Services, 2006[15]

The Arcadis/RPA (2008)[13] report for the European Commission analysed the impact of the CPR systems established by the WEEE Directive in Europe on product design. The report stated that evidence that the WEEE Directive has provided incentives for eco-design is inconclusive. This demonstrates that CPR does not provide an incentive to a producer to design products to be easier to recycle.

4.4.4 Orphan Waste and Free-riding

Products deposited for recycling that are the responsibility of a company that is either no longer present in the market or has not paid for its recycling is known as orphan waste. The producers responsible for orphan waste are known as free-riders. High amounts of orphan waste create problems for WEEE recycling systems as these costs need to be covered by the remaining producers.

The ICT Milieu return-share IPR system was criticised for resulting in a high level of orphan waste. In 2002, 35% of all equipment collected was orphan or free-rider products. As a result, the system was changed for 2003. However, despite moving to a market-share-based system, according to recent samples by ICT Milieu orphan waste remains at 20–25% in the Netherlands.[14]

Data on the estimated levels of free-riders in European CPR schemes in 2006 show that free-riders currently represent between 10 and 20% by volume of products placed on the market.[15]

In contrast, orphaned products constitute roughly 5% of the recycled products in Japan. In Maine, whose return-share IPR system is closely comparable to ICT Milieu, orphan waste constitutes 4.8% of the total volume of electronic waste. This lower figure is attributable to stronger enforcement through banning the sale of brands that are not registered to a producer that is compliant with the producer-responsibility law. In Europe many producers advocate a similar system of enforcement, where legislation only allows products to be sold where their producers could provide proof of registration.

5   Recommendations to Implement IPR

This section outlines a series of recommendations in order to implement IPR within the European Union.

5.1   Recommendation #1: Ensure Article 8.2 of the WEEE Directive is Fully Transposed

The first element that needs attention is the lack of implementation of the concept of IPR through the European Member States' national WEEE laws. The WEEE Directive itself encompasses the concept of IPR through Article 8.2. In several EU member states this article has been ignored.

5.2   Recommendation #2: Adopt a Phased Approach to IPR

While recognising that CPR is appropriate for handling historic waste (*i.e.* waste that was placed on the market prior to 13 August 2005), producers should be able to move towards a system of IPR, whereby each producer will be liable for financing the treatment and recycling of their own waste products.

IPR needs to be introduced in a staged manner. In order to make the transition from collective responsibility, worked out according to market share for historic waste, to handling future waste under a system of IPR, the most logical step would be to first move from collective responsibility to a system using 'return share'. Under a system of return share, responsibility for waste would be based on the producer's proportion of the actual waste returned and would, therefore, be a truer representation of a producer's waste responsibility.

This transition stage is necessary as full IPR will involve the sorting of waste so it can be returned to the original producer. Sorting all WEEE is considered to be disproportionately capital- and labour-intensive, but future design and recycling technologies may address this sorting issue in the future.

The transition from a fully collective system to a system where producers can comply through IPR is further described and shown in Figure 4.

5.2.1   Phase 1: Fully Collective Responsibility
This implementation model is described in Section 2.2. All producers are responsible for a proportion of the WEEE returned according to their current

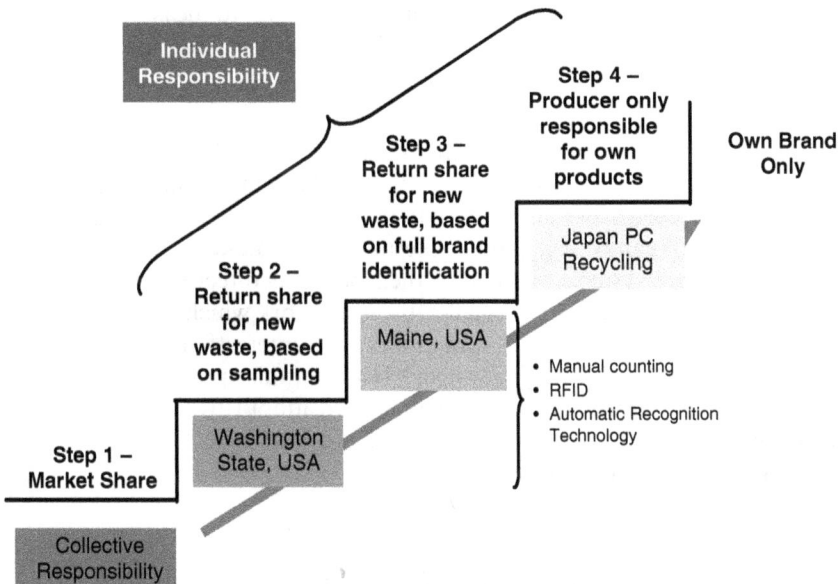

Figure 4   A phased approach to the implementation of individual producer responsibility.

market share. This system is the most commonly implemented one across Europe and in many other countries.

5.2.2   Phase 2: Return-Share System Based on Sampling

This model is used in Washington State, USA, and is described in Section 3.2. In order to effect the transition from collective responsibility for historic waste to handling future waste under an own-brand or individual system, the most logical step is to first move to a system using a 'return-share' allocation. Under a system of return-share allocation, responsibility for waste is based on the producer's proportion of the actual waste returned, and not the proportion of EEE it is currently placing on the market. It is, therefore, a truer representation of a producer's waste responsibility than current market-share.

In Phase 2, return share is based upon sampling of the waste stream. Random sampling of the collection containers enables the proportion of each brand manufacturer's waste in each type of WEEE waste stream to be calculated. These data are known as protocols. For those producers who wish to adopt IPR, these protocols could then be used rather than market share.

5.2.3   Phase 3: Return-Share System Based on Product Identification

Phase 3 extends the 'return-share' allocation by replacing the sampling approach with a system which is based upon product identification. Rather than protocols being established based upon samples from collection containers, these protocols are established from an exact measurement of the numbers of products in the waste stream. Technologies to measure products in the waste stream are emerging. These include systems based upon bar-codes and Radio Frequency Identification (RFID) tags.

5.2.4   Phase 4: Full IPR Based on Brand Responsibility

The final phase is that of full IPR. Full IPR can be achieved in two ways. First using the technologies identified above to identify products in the waste stream, individual brands can be segregated. Alternatively, producers can build on their existing take-back systems and collect their products directly from end users.

This phased approach provides a flexible framework which can suit the needs of different types of companies and protect the interests of companies who prefer to remain within collective systems. However this approach also provides the opportunity for those companies who wish to adopt IPR to be able to do so. There should be freedom of choice whether producers use IPR or not as this will be a commercial decision.

5.3   Recommendation #3: Member States to Implement IPR

This phased approach needs to be accompanied by an implementation strategy by each Member State. The following is an indicative list of the actions needed by each Member State to enable IPR:

- Member State guidance needs to be amended to recognise the split between historic and future WEEE, and to enable producers to take an IPR for future WEEE.
- Guidance needs to determine the agreed sampling methodology for compliance schemes, to determine producers' return-share obligations and to determine the approach of the Member State towards financial guarantees and grey imports (grey imports are products imported by entrepreneurs exploiting the lower price of a product elsewhere in the world).
- Government/Clearing House develop a brand-responsibility spreadsheet per country to determine the producers relevant to each brand in the waste stream.
- Compliance schemes to sample waste stream according to agreed methodology and report data to Government or Clearing House.
- Member State or Clearing House can then determine return share obligations for producers.
- Compliance schemes will issue invoices to producers based on their return share.

6 Conclusions

This chapter has described the background to e-waste legislation and reviewed the common approaches to producer responsibility for e-waste, including individual *versus* collective producer responsibility and competitive and non-competitive systems of compliance.

The chapter explains the distinction between financial responsibility and the operation recycling systems which are used to provide compliance. The implementation of Producer Responsibility illustrates that environmental benefits and economic costs vary widely, dependent upon the approaches adopted by individual nation states.

The chapter has revealed that many countries have failed to transpose the IPR concept into their national laws and instead implemented CPR for all waste.

By examining a number of examples of IPR in practice, it can be seen that IPR is a viable policy option. It has been adopted in Japan and the USA, and was operating in the Netherlands until 2003. Evaluating IPR against CPR demonstrates that both approaches can deliver comparable levels of collection and recycling. However, CPR does not provide incentives for producers to improve the design of their products whereas IPR in Japan has led to significant improvements in the design of products.

At the operational level, the chapter reviews the performance of competitive systems *versus* non-competitive systems. Evidence demonstrates that competitive recycling systems lead to significantly lower costs of compliance.

References

1. J. Huisman, R. Kuehr, F. Magalini, S. Ogilvie, C. Maurer, E. Artim, C. Delgado and A. Stevels, *2008 Review of Directive 2002/96 on Waste Electrical and Electronic Equipment (WEEE) – Final Report* (United Nations University, Germany), 2007.
2. J. Pichtel, *Waste Management Practices: Municipal, Hazardous, and Industrial,* Taylor & Francis Group, Boca Raton, Florida, 2005.
3. C. Van Rossem, N. Tojo, and T. Lindhqvist, *Lost in Transposition?: A Study of the Implementation of Individual Producer Responsibility in the WEEE Directive*, Greenpeace International, Friends of the Earth Europe and The European Environmental Bureau, 2006.
4. Directive 2002/96/EC of the European Parliament and of the Council of 27 January 2003 on the waste electrical and electronic equipment (WEEE), Official Journal of the European Union, 13/02/2003.
5. K. Mayers, C. M. France and S. J. Cowell, *J. Ind. Ecol.*, 2005, **9**(3), 169–189.
6. Proposal for a Directive of the European Parliament and of the Council on Waste Electrical and Electronic Equipment, Proposal for a Directive of the European Parliament and of the Council on the Restriction of the Use of Certain Hazardous Substances in Electrical and Electronic Equipment, COM (2000) 347 final.
7. IPR Works, *Individual Producer Responsibility Works*, 2007, available from http://www.IPRworks.org, accessed 25/03/08.
8. K. Sander, S. Schilling, N. Tojo, C. van Rossem, J. Verson and C. George, *The Producer Responsibility Principle of the WEEE Directive Final Report,* Okopol, Germany, 2007.
9. Maine State Legislature, *Statute Title 38, Chapter 16: Sale of Consumer Products Affecting the Environment*, available from http://janus.state.me.us/legis/, 2008, accessed 25/03/08.
10. EPHC (Environmental Protection and Heritage Council), *Industry Discussion Paper on Co-regulatory Frameworks for Product Stewardship*, EPHC, Australia, 2005.
11. R. Veit, How do WEEE get it right?, available from http://www.emsnow.com/npps/story.cfm?id=15184/, 2005, accessed 25/03/08.
12. N. Tojo, EPR program for EEE in Japan: Brand Separation?, in *INSEAD WEEE Directive Series*, 2006.
13. R. P. A. Arcadis, *WEEE component – The impacts of the WEEE Directive and its requirements with respect to various aspects of innovation and competition – Draft Report*, Arcadis, RPA, Belgium, 2007.
14. J. Vlak, Personal communication, 15 February, 2008.
15. Bio Intelligence Service, *Gather, Process, and Summarise Information for the Review of the Waste Electric and Electronic Equipment Directive (2002/96/EC) Synthesis report final version*, Bio Intelligence Service, France, 2006.
16. Hewlett Packard, *Real Consumer Costs for Electronic Equipment Recycling as Low as 1 Euro Cent*, Press release, issued 5 June 2006.

17. World Development Indicator (WDI) reports, World Bank Publications, 2000.
18. World Development Indicator (WDI) reports, World Bank Publications, 2006.
19. Little Data Book, WDI, World Bank Publications, 2003.
20. Little Data Book, WDI, World Bank Publications, 2004.
21. Little Data Book, WDI, World Bank Publications, 2005.
22. K. McIntyre, *Mandatory Take Back: 2010*, 2007, unpublished.

Rapid Assessment of Electronics Enclosure Plastics

PATRICK J. BAIRD, HENRYK HERMAN AND GARY C. STEVENS

1 Introduction

Current recycling qualification methods for plastic materials such as poly-carbonate (PC), acrylonitrile butadiene styrene (ABS), PC/ABS blends, high-impact polystyrene (HIPS) and polyphenylene oxide (PPO) are inefficient and any recycling that is performed generally produces relatively poorly qualified and at times poorly identified materials which are unable to achieve their ultimate recyclate market value. In many cases even simple identification of single materials or mixed material characterisation is not attempted and only low-value mixed recyclate is produced which is only fit for low-value applications. The driver for developing methods to retrieve more value and use from these high-quality materials at end of life is now supported by key European producer-responsibility legislation such as the Waste Electrical and Electronic Equipment (WEEE) directive[1] which aims to promote the recovery and re-use of materials arising from the electronics sector. Further discussion of the drivers is presented in Chapter 2: 'Materials Used in Manufacturing Electrical and Electronic Products' by Stevens and Goosey.

The development of tools to aid in the identification, separation and qualification of engineering thermoplastics is urgently required to extract the maximum economic and resource utilisation value from these materials and to comply with new legislative requirements, as discussed in Chapter 2. These tools need to be capable of rapidly and reliably identifying plastics.[2,3] Where possible they should be hand-held, robust and portable yet readily incorporated into conveyor-belt systems suitable for automated analysis on a disassembly or materials-separation line. Other requirements that are important include the

Issues in Environmental Science and Technology, 27
Electronic Waste Management
Edited by R.E. Hester and R.M. Harrison
© Royal Society of Chemistry 2009
Published by the Royal Society of Chemistry, www.rsc.org

ability to deal with darker plastics containing carbon black and other additives, such as flame-retardants and anti-oxidants. Such tools must, however, not be prohibitively expensive and they must be versatile and useful in different parts of the product lifecycle (for instance the qualification of virgin material and the qualification of materials from product disassembly or post-shredder separation).

In addition to the simple differentiation of key plastic types, the ability to qualify these materials for re-use at any stage of the product lifecycle would be very attractive. The capability of determining material properties, processability and expected remaining life or usefulness in particular applications would enable recyclates to be certified for resale, giving component manufacturers the confidence to re-use these materials in production lines. Even the value of mixed waste streams could be increased if certified to meet specified minimum requirements.

Many of these factors are appropriate for rapid assessment by molecular spectroscopy. Colour is an obvious discriminant when the visible part of the spectrum is used, but molecular structure, modulated by packing, density and crystallinity provides spectral information in the near and far infrared, resulting from the fundamental vibrations of the molecules that are more useful for predictive analysis. The application criteria mentioned above limit the spectral range, so the mid-infrared (wavelength 2.5 to 25 microns) is less desirable. In contrast, the combined visible-near infrared (VIS-NIR) spectral range of 350 to 2500 nm is very useful and this has been used in a new portable spectrometer developed at the University of Surrey and GnoSys UK Ltd.[4] This system also uses multivariate statistical analysis (MVSA)[5] to rapidly determine correlations between the spectroscopic measurements and various material properties and provides a powerful and versatile instrument for this type of application.

The outputs from these new developments are providing valuable information that is being fed back into the design of efficient online measurement systems.[6,7] In this chapter we describe the work done in identifying and discriminating between polymer types, the online determination of physical properties that can be related to simulated ageing of these materials in the laboratory and the analysis of additives in plastics.[8]

2 Instrumental Techniques

An exemplar of a portable measurement system (TRANSPEC™)[9] that was developed at GnoSys Ltd and the University of Surrey for use in the power industry, consists of integrated miniature spectrometers covering the visible (VIS) to near-infrared (NIR) range (350–2500 nm wavelength) using diffuse reflectance probes with fibre-optic connections and electronic control interfacing – this is shown schematically in Figure 1. The system is readily applicable to measurement of plastics and other organic materials. The instrument combines diffuse reflectance spectroscopy with MVSA, operating

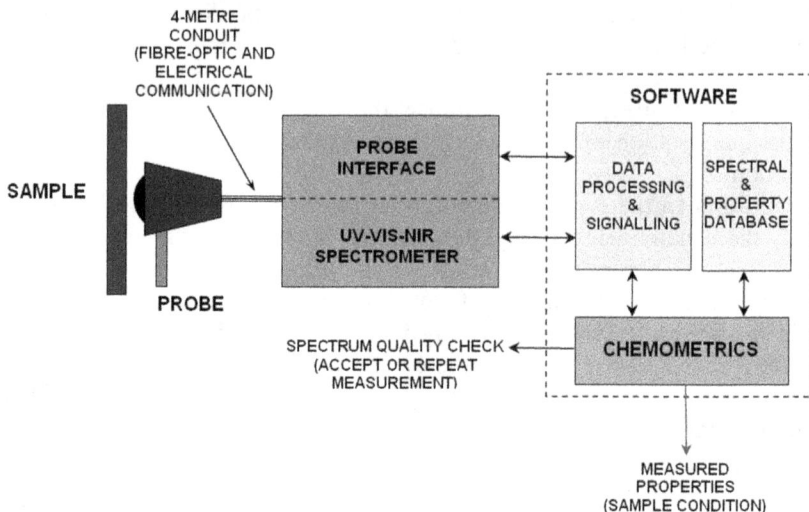

Figure 1   TRANSPEC spectroscopic measurement system.

within a newly developed software environment (TRANSCHEM™) (Figure 2), to provide a wide-wavelength spectroscopic analyser for rapid online condition assessment of materials. The system is modular and can be extended to accommodate a range of probes and interfacing adapted for different measurement applications.

MVSA software is a valuable data analysis tool that is used to analyse the spectra and provide information relating to physical and chemical properties. Using methods such as principal components analysis (PCA), regression (PCR) and partial least squares (PLS), discrimination of sample groups and correlations between spectra and sample properties can be determined. Calibration models from a sample archive are used to predict the same properties from samples measured online, providing the new sample property values occur within the range spanned by the statistical models. The archives are periodically expanded and, with enough samples in the database to cover the desired range and variance, reliable models are generated for typical *in-situ* measurement of sample properties in the field.

Other variants of this method use infrared and Raman spectroscopic techniques to access molecular and material data, if required, with much higher spatial resolution. These methods are useful for investigating some specific types of polymer additive that are not readily determined by visible-NIR spectroscopy because of the larger molecular groups involved.

Other techniques used are X-ray fluorescence (XRF) for elemental analysis[10] and an optical emission spectroscopy (OES) method called Sliding-Spark spectroscopy,[11] both of which provide valuable supplementary information relating to polymer additives and also serve as an alternative method of

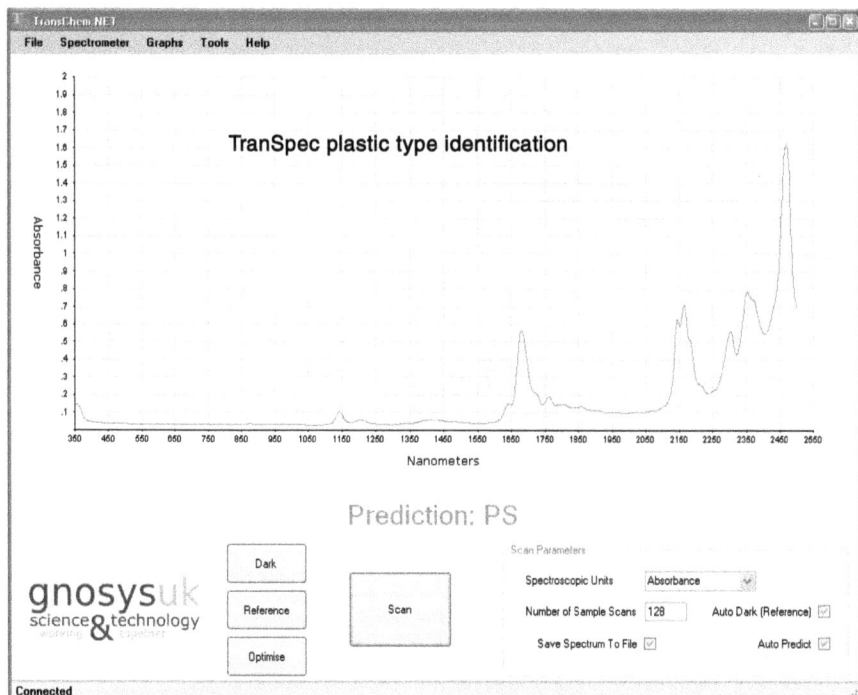

Figure 2 Online identification of plastic type with TRANSPEC.

measurement in the case of black materials which are not so easily analysed by molecular spectroscopy. Other possibilities for the vibrational spectroscopy of materials containing carbon black, such as portable mid-infrared or micro-Raman, have been reported elsewhere.[12–14]

Various methods are used to generate the database of sample properties required for the correlation with spectral information that is used in the spectrometer calibration models. In the case of base polymers a list of physical and chemical properties is often available. For processed polymers, parameters such as melt flow rate (MFR), tensile strength and molecular distribution are measured using standard techniques. In the case of polymers containing additives, data on added materials such as flame retardants (FRs) and anti-oxidants can also be obtained or formulated as stable standard samples.

3 Visible-NIR Spectroscopy of Engineering Thermoplastics

The ability to differentiate polymers using optical spectroscopy is well established, and NIR spectroscopy (800 to 2500 nm wavelength) is particularly good at discriminating polymer types. A combination of VIS (350 to 800 nm) and NIR can help by using colour information to inform the NIR data.

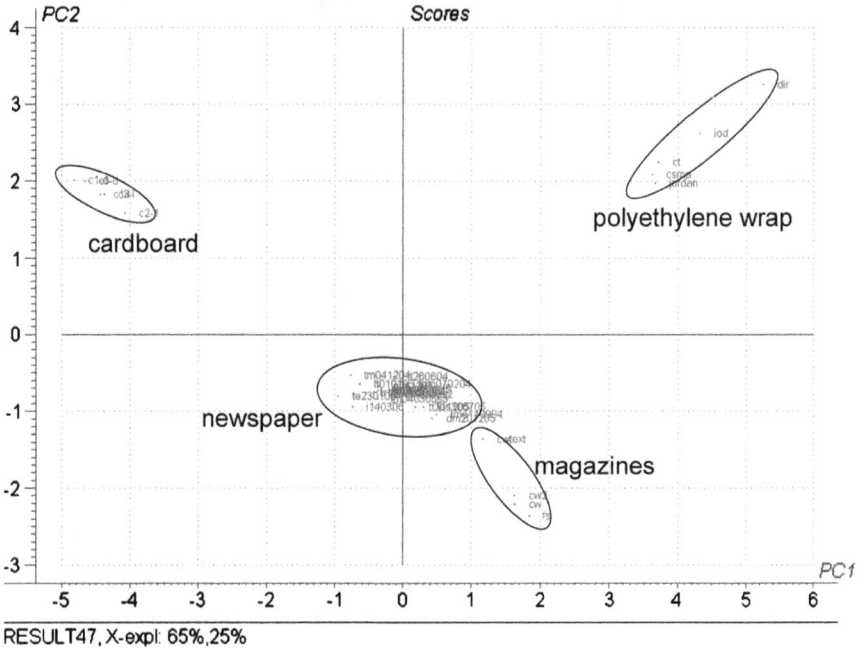

Figure 3   PCA of different recyclate materials.

The method can be used in a more global sense to separate different materials in waste streams. Clear separation of plastic materials from paper and other materials can be seen, as for example the variation in spectral information shown by the cluster analysis in Figure 3. The same methods can be adopted at an earlier stage of separation before the separation of individual plastic types is required.

Principal Component Analysis (PCA) is the method used to generate the cluster analysis and this can be applied at many different levels, depending on user needs. In this type of analysis, each spectrum is reduced to a point in a complex statistical space, with similar materials being grouped together. PCA acts on the matrix of spectroscopic data only (ignoring any sample property information that may be available) and reduces the variables (in this case wavelengths) to a small number of linearly independent principal components (PCs) that describe the majority of the variance across the samples, thus picking out the trends in the data and separating them from background noise. The spectroscopic data matrix is effectively decomposed into two smaller matrices, the 'scores' matrix (the new PCs) and the 'loadings' matrix (the transformations). In this way each PC is described by a weighted combination of values relating to each wavelength. Some wavelengths will have much more influence on the PCs than others. The scores matrix contains the new values or coordinates of the data objects in PC space. These are simplified versions of the original spectral intensities. Similar samples will group or cluster together in PC

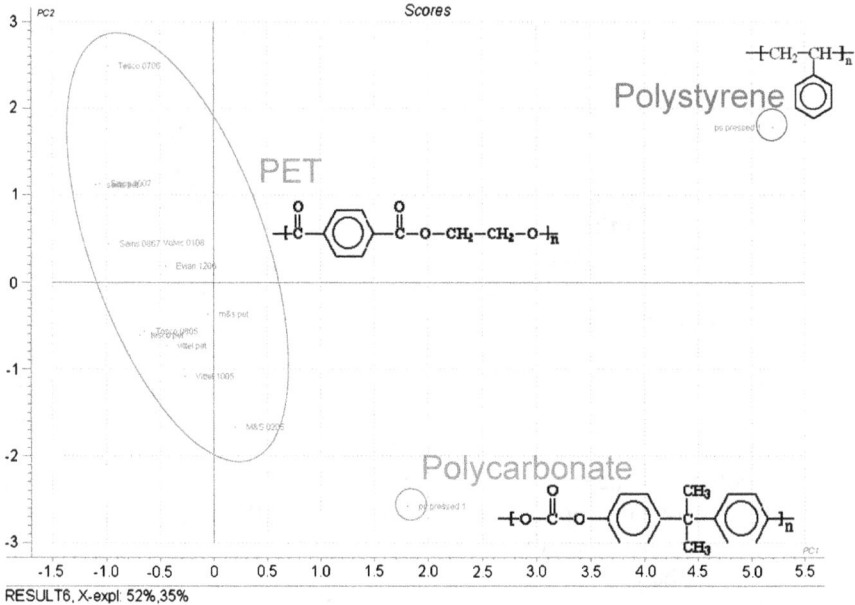

Figure 4 PCA cluster plot of three polymers.

space and graphical representation of the scores in PC space provides a method for sample identification.

Comparing some polyethylene terephthalate (PET) samples from used bottles with similar unpigmented samples of polystyrene and polycarbonate using PCA shows how strongly and easily these polymer types are discriminated (see Figure 4).

These methods are capable of discriminating materials, but it is also possible to 'qualify' them by predicting the physical properties and processability of the material using multivariate statistical analysis, as described in Section 3.5.

3.1 Discrimination of Enclosure Materials

An extension of the PCA methodology, called Soft Independent Modelling of Class Analogies (SIMCA), allows for at least two levels of discrimination and property prediction: the first sorts samples into 'types', and the second includes a model for each 'type' that predicts the properties, as shown later. Since this initial analysis, discriminant models have been developed that include many polymer types, blends and sample geometries, and installed on our equipment to serve as a rapid online classification method for thermoplastics and polymeric materials in general. If raw spectra are used for the analysis, problems can arise due to the variation in sample reflectivity that depends on sample thickness, shape, surface texture and curvature, particle size, *etc*. This was observed in our original data sets. However, following the transformation of spectra using pre-processing methods such as multiple scatter correction (MSC) and derivative calculations, the non-specific variance can be much reduced and the resultant models are able to identify materials regardless of their physical appearance (the models are able to discriminate chemical content regardless of whether or not the samples are, for example, in the form of powders, pellets, thin strips or plaques). Precise colour index measurement of the materials can also be determined in the same measurement.

The plastics can be identified even in the presence of flame retardants. Different pre-processing methods are required for identification ('discriminant') models of plastic type. The example in Figure 5 is another PCA scores plot (or mapping) which shows that the different polymer types are clustered into distinct groups and this clustering can be used to determine the plastic type of a new measurement, provided that this plastic type is represented in the model.

Seven plastic types are represented in this data set: ABS, HIPS, PS, PC, PC/ABS, polybutadiene (PB) and PET. In Figure 5 it appears that ABS and HIPS are clustered together, and this is because ABS contains polystyrene which accounts for much of the spectral information due to the strong absorbances of the aromatic components. However, if we look closer at the ABS and HIPS cluster we can see that the ABS and HIPS samples can be separated and we can also distinguish between different origins or grades

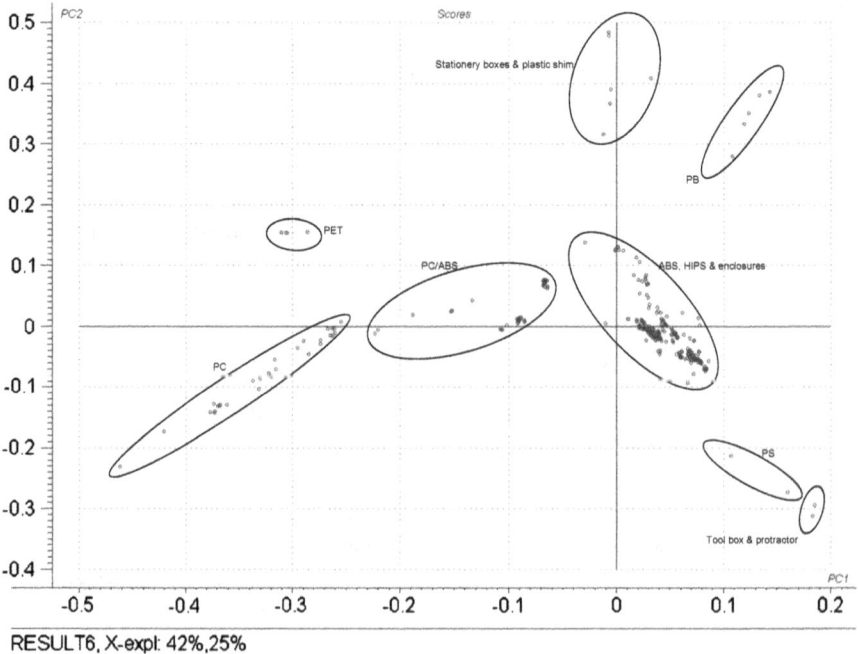

Figure 5 PCA cluster plot of plastic types.

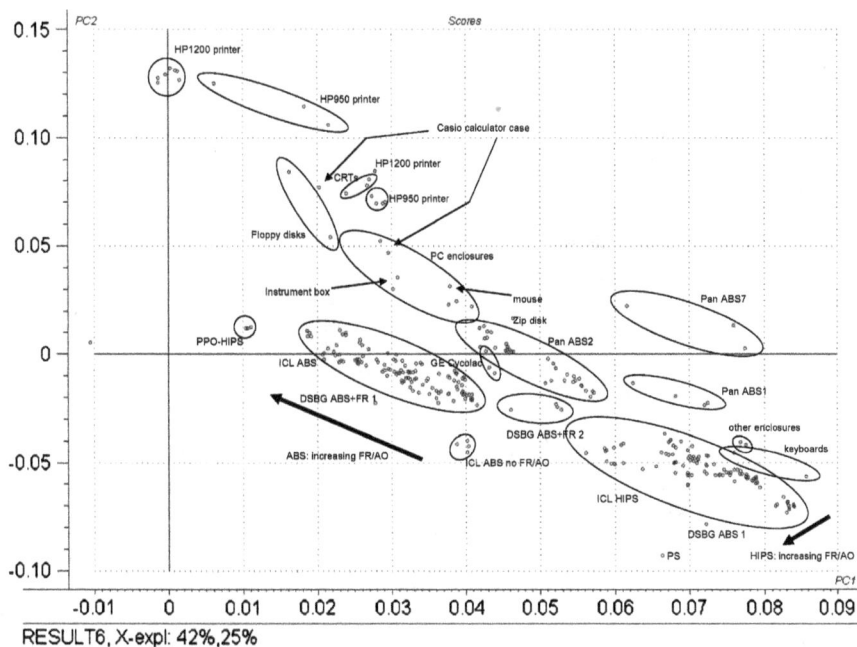

Figure 6 PCA cluster plot in more detail showing separate groups of ABS and HIPS.

of ABS, as displayed in Figure 6. Measurements of plastic enclosures from various office/IT/laboratory items are included in the plot and from this it can be estimated which area cluster or type of material they are closest to. Discriminant models such as these are stored with our instrument control software for use in online identification. The models work well for WEEE materials (electronic enclosure waste) provided the appropriate transformations are applied to the spectra – this has been tested on a large stock of waste products that have been acquired from industry and shown to be robust. Plastic type identification results can be displayed immediately following a scan.

3.2 Base Polymer Identification

For the purpose of assigning specific molecular group vibrations to spectral features we acquired and examined the three main contributors to electronic equipment enclosures: ABS, PC and PC/ABS blend, including the FR grades. Figure 7 shows spectra of the four polymers polystyrene (PS), polyacrylonitrile (PAN), polybutadiene rubber (PB) and polycarbonate (PC); these four simple polymers represent all the molecular groups to be found in the more complex polymers under investigation. Some information on spectral features of these polymers can be found in the literature[15-19] but this is limited; some of the spectral assignments, which represent vibrational overtones and combinations,

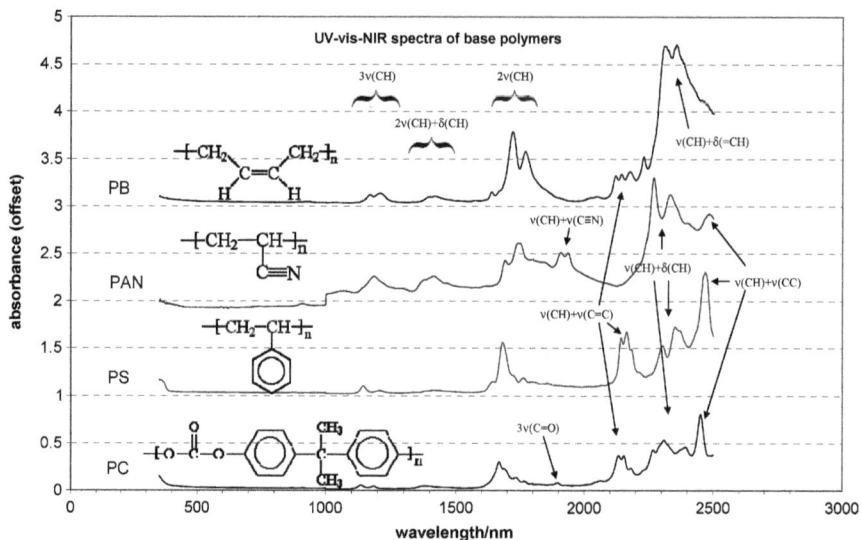

Figure 7   Spectral features of base polymers.

are therefore inferred from infrared data. The main groups of these are displayed for the prominent spectral features or bands that appear in the NIR: in Figure 7, molecular bands are shown as 'v' (bond stretching) and 'δ' (deformation).

It is important to note that while it is common practice to use such bands discretely in materials identification, such bands are subtly altered by their local chemical and physical environment and those changes can be captured effectively in 'whole' spectral analysis methods. Bands are typically broad in the NIR due to all the possible combinations of fundamental molecular vibrations in the infrared and are sensitive to the interactions between molecules, and environmental effects can be subtle. Additives will also contribute and these can be identified by their contribution to the 'whole' spectral response.

Features in the spectra shown in Figure 7 appear also in the spectra of the copolymers, with intensities according to the relative concentration of these components in ABS and PC/ABS. Similar features appear in polycarbonate with an additional peak at around 1900 nm.

3.3   Selected Thermoplastics for Processing

Spectroscopic measurements of ABS, PC and PC/ABS samples have shown that there is a clean discrimination between these three polymer types. For the purpose of studying the effects of controlled degradation, ABS and PC/ABS polymers were selected due to the abundance of these two materials in WEEE.

3.4 Controlled Degradation Experiments

The effect of polymer processing and weathering on physical properties has been reported in the literature.[20–22] For investigation of these effects we subjected ABS and PC/ABS samples to repeat processing through ten passes of injection moulding, retaining samples at each stage;[7] following the first pass of injection moulding, samples of the same polymers were also subjected to 500 hours of controlled UV exposure/spray weathering (UV-B in batches of 6 hours exposure, 2 hours condensing), with samples being retained at roughly 100-hour intervals for mechanical and spectroscopic examination. The injection moulding process and requirement for sample retention at various stages required an initial 20 kg of raw polymer material. The samples were obtained as dumb-bell-shaped plaques of thickness 4 mm. The samples were measured spectroscopically with a white 100% reflectance tile behind them, averaging over several spectra per sample.

3.5 Analysis of Processed Thermoplastics

The value of the recyclate polymers is related to the mechanical and thermal properties of the materials and their intrinsic market economic value. Although there will have to be some averaging if the batch contains a number of sources, and a metric will have to be employed to compute an 'average' value of the property, a rapid way of measuring a large fraction of that batch will at least provide a property/volume fraction parameter.

It is clear from our findings that progressive repeat injection moulding of the materials results in a progressive deterioration in condition. UV ageing also results in changes, but with a somewhat different trajectory, as observed in the PCA plots shown for ABS (Figure 8) and PC/ABS (Figure 9). This in turn gives us information about the *type* and *process* of degradation.

To correlate the spectral information with sample properties, the PCA process can be extended to Principal Components Regression (PCR) or Partial Least Squares (PLS). These processes undergo matrix decomposition as before, but take into account the known (calibration) property data as well. In the case of PCR, regression is applied between the resulting matrix from PCA and the known property data to create a model that can be used for predictions on similar samples. In the case of PLS, the property data are used in computing the PCs themselves, and therefore the result depends on the accuracy of the property data as well as the spectral information, but this is a preferred method if the property data are a weak function of the spectra. Once the models are constructed, they can be used to predict the sample property of newly measured samples, provided that the property value falls within the range defined by the calibration.

PLS models were generated from spectra and property data measured at regular intervals between processing and weathering of the plastics. By way of example, we show that there is a strong relationship between UV exposure and the VIS-NIR spectra. Figure 10 shows that over this exposure period there is an

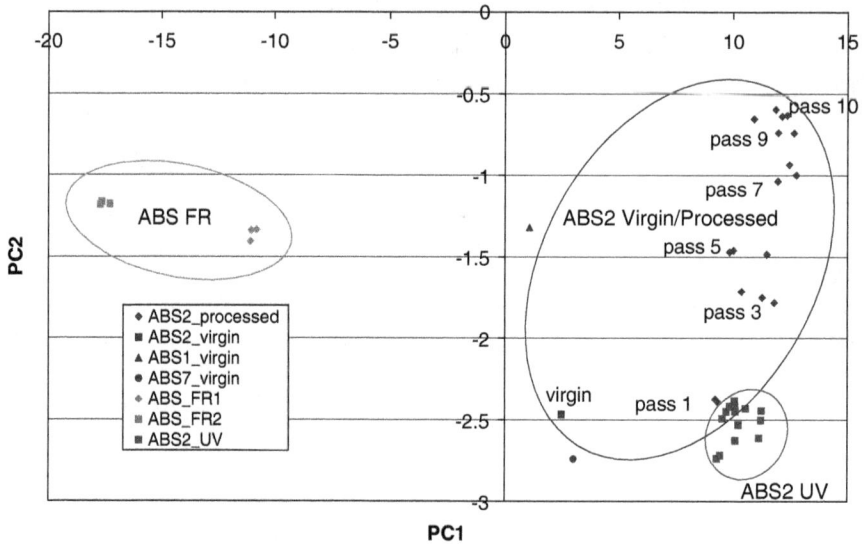

Figure 8   PCA cluster plot of processed/weathered ABS.

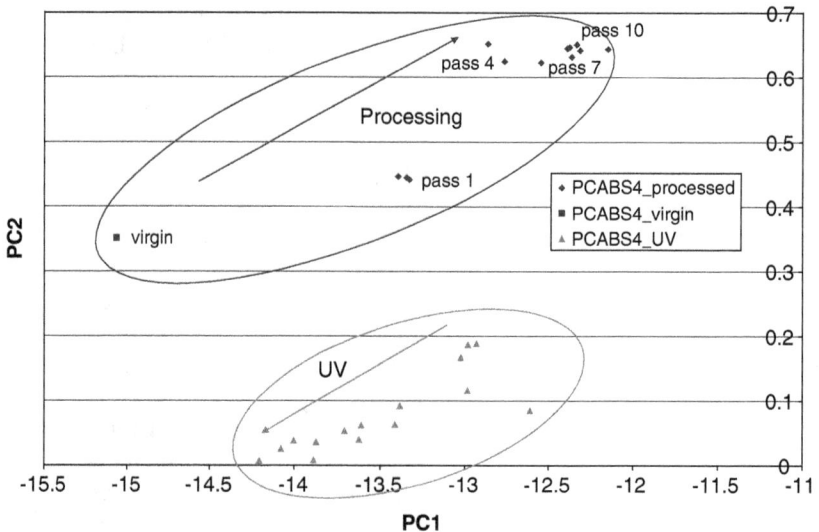

Figure 9   PCA cluster plot of processed/weathered PC/ABS.

excellent correlation of 0.98, with standard error of prediction of about 32 hours (2σ). The correlation is mainly through colour changes – an increasing absorbance in the blue, but tempered by transfer of UV/blue absorbing species into the green part of the visible spectrum, all accompanied by changes in the level of carbon unsaturation.

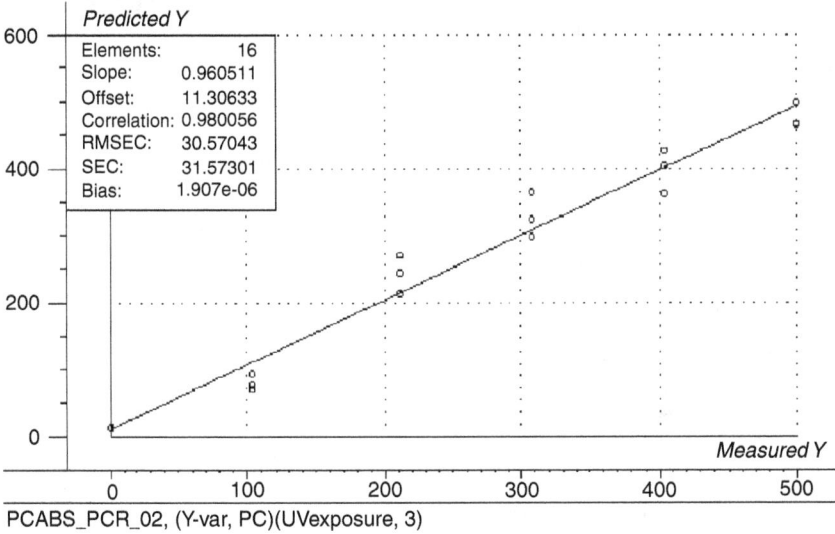

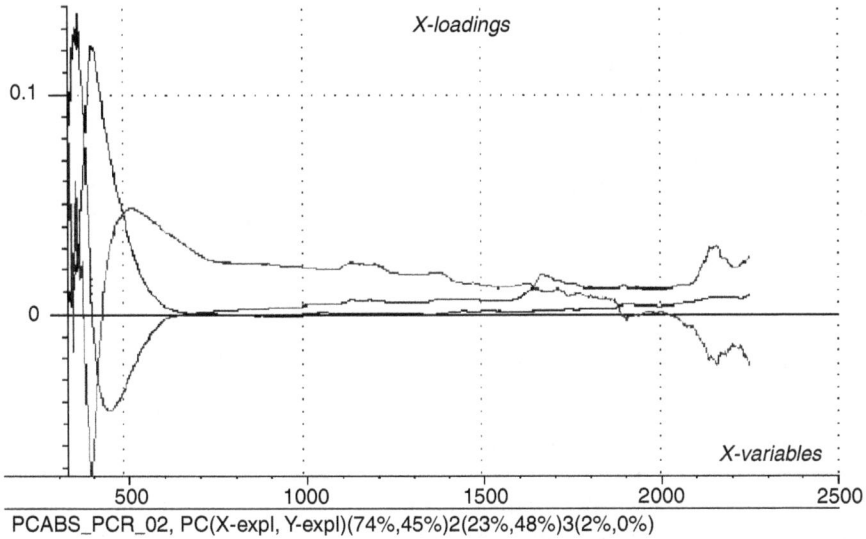

Figure 10 PLS analysis of UV exposure for PC/ABS.

This same dataset can be used to predict the tensile modulus and the melt flow rate (MFR) for the PC/ABS polymer system. UV ageing affects some properties, in particular the tensile modulus. The MFR is also affected by UV, but in the first 100 hrs only, and to a lesser extent. This suggests that the UV ageing is affecting mainly the surface, and not the bulk of the material that is used for the MFR measurement.[23]

It is worth noting that in the case of ABS, the first stage of UV exposure (100 hrs) causes an increase in MFR whilst for the PC/ABS blend a similar UV exposure has a lesser effect. This suggests that in ABS the UV causes bond breaking in the surface layers, probably the unsaturated carbon bonds (C=C) in the butadiene component,[23] which then cross-link further, thus increasing the tensile strength. This observation is complemented by the effect of UV exposure on stress at break which is seen in ABS only; in PC/ABS there is no observable change in this parameter. This suggests that in PC/ABS the carbonyl (C=O) component of the PC is absorbing much of the UV radiation and the resulting change in physical properties is less than in the butadiene component in ABS.

The results for the injection-moulded polymers suggested that PC/ABS is significantly more affected by mechanical processing than ABS, as suggested by the changes observed in all parameters between one pass and the next. Significant increase in tensile modulus and MFR is accompanied by decrease in the other four yield/break parameters, so it is clear that the PC component is more affected by processing than ABS with each injection-mould pass. A model made from the restricted data set available currently indicates MFR predictions accurate to 1 g/10 min over the range 7 to 34 g/10 min.

To validate some of these experiments we also used Gel Permeation Chromatography (GPC) to determine how the molecular weight distribution was affected by the different types of processing in the two copolymers. In the case of ABS, UV weathering had no observable effect at all, whereas the molecular weight distribution decreased with increased mechanical processing. In the case of PC/ABS, there was a small change in molecular weight distribution due to mechanical processing, observed beyond pass 7, and a slight change resulting from UV weathering.

4　Analysis of Plastics Containing Flame-retardant Additives

The same process can be applied to the identification of flame-retardant (FR) content in the polymer, specifically those flame-retardant materials that have some organic content, which is generally required for NIR analysis. The existence of other types of flame-retardant content may be inferred from UV-visible spectroscopy as a result of colour change but a more appropriate method, such as infrared or Raman spectroscopy, is required to accurately identify the chemical type in these cases. Our UV/VIS/NIR measurements using our analysis software have been supplemented by infrared and Raman measurements on some of the same samples, and in addition to this we performed XRF and OES for an accurate determination of bromine (Br) content and identification of other halogens and metals in the material. We acquired plaque samples of ABS, PC/ABS and HIPS with and without various known concentrations of a number of different brominated flame retardants, in addition to varying amounts of antimony trioxide as a co-synergist. Varying smaller quantities of other additives such as antioxidants, stabilisers and modifiers were expected to be present, and XRF data indicated the existence of further additives.

4.1 Visible-NIR Spectroscopy

The presence of the antimony trioxide additive has the effect of increasing the reflectivity of the plastic surface, and this is accounted for in the spectral analysis. In the case of FRs which contain no molecular groups involving hydrogen, such as decabromodiphenyl ether (decaBDE), there are no detectable overtones in the NIR range, and the only differences observed are colour changes in the visible region approaching the UV. If required, the colour differences can be used (using the absorbance data) to produce conventional colour index values, which could be useful for determining the presence of FR in the plastic; however, these types of FR require infrared or Raman spectroscopy for proper identification. Some of the FRs contain CH groups but comparatively a lot of bromine so the concentration of organic material is low, and there are examples of CH groups in all three components of ABS so no significant additional variation is observed in these cases either, except for slight wavelength shifts in the NIR peaks, typically relating to unsaturated CH combinations in the polystyrene component of the polymer. In the case of FRs containing OH groups, such as those in tetrabromobisphenol A (TBBA) and the more recent brominated epoxy resin types, there are some clear differences observed. In the case of TBBA these relate to the OH groups attached to the aromatic rings, which show strong features in the NIR (Figure 11). In the case of the epoxy FRs, we can see distinct features due to OH occurring at a different wavelength because the OH group is attached to a chain link rather than a ring.

Spectral data are transformed to minimise the effect of reflectance variation. MVSA is used again to reduce the spectra to a simpler form and predict the FR content. PCA is first applied to the data, in which it is shown that the data are readily separated into plastic type regardless of their FR content (Figure 12).

Following this sample-type separation, PCA can again be performed for a particular plastic type, *e.g.* ABS, in which the general trend of increasing FR content can be seen, and the different FRs are grouped in distinct data populations (Figure 13). The two epoxy-type FRs sit in similar locations compared with the TBBA, but are still distinct from one another.

PLS models were generated for those flame retardants that showed variation in the spectra. The polymer information in the spectra is much better represented than the FR content, so care is required not simply to represent the amount of polymer in the regression model, because the spectral intensities are related to material concentration, and increasing FR content corresponds with a decrease in polymer content. Selection of regions of the spectra relating

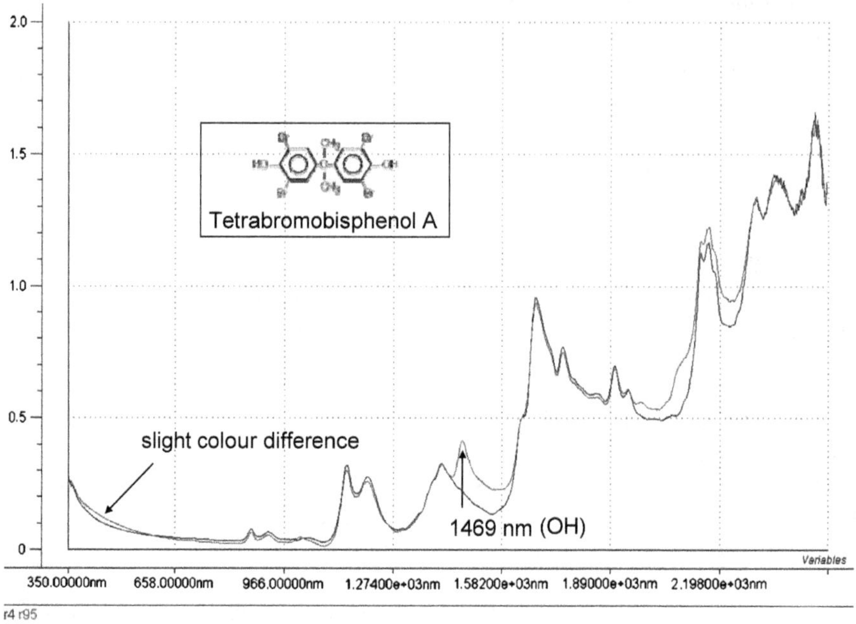

Figure 11   ABS with and without TBBA.

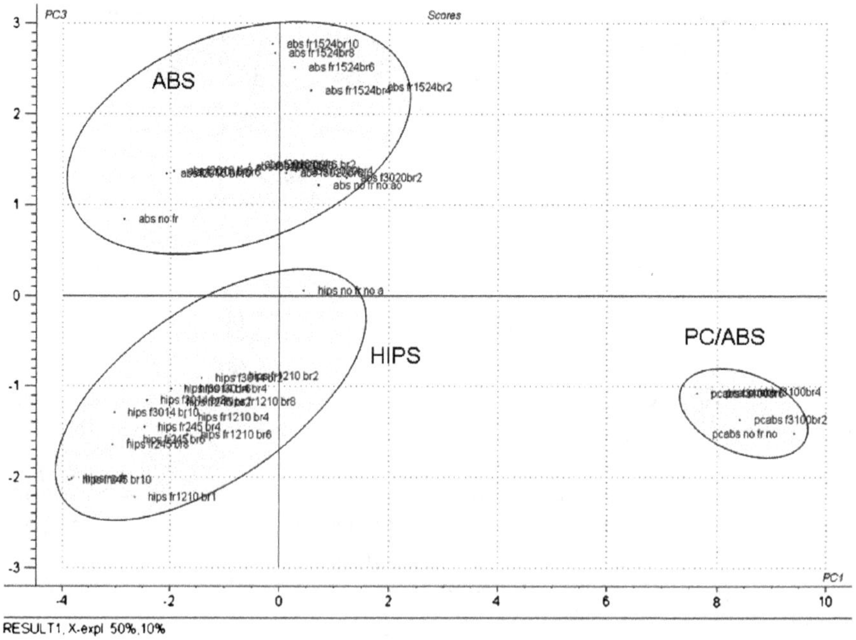

Figure 12   PCA cluster plot showing discrimination between polymer types.

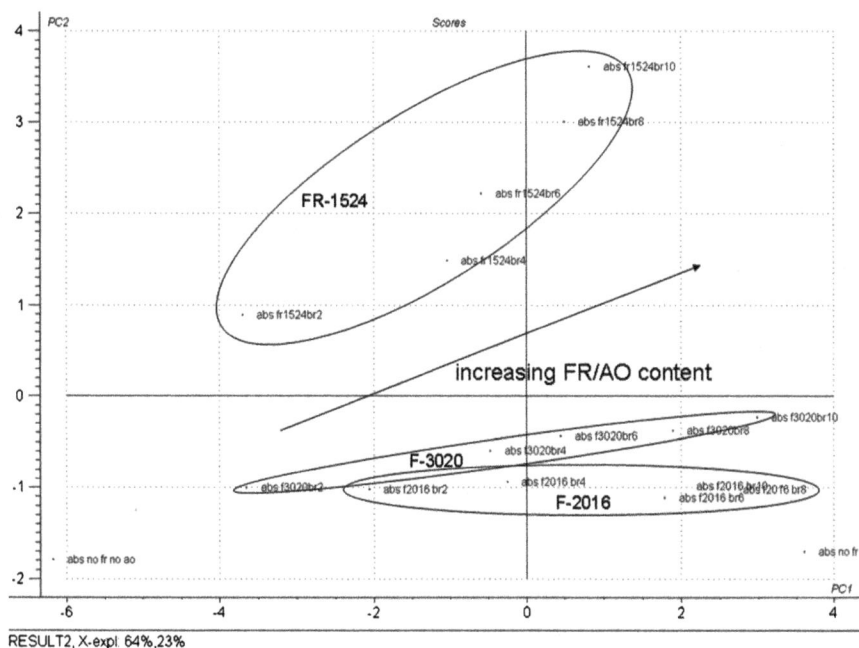

Figure 13 PCA cluster plot for ABS showing trends relating to FR types and concentrations.

to FR content only can help to minimise this problem. Accurate models were obtainable for half of the FRs investigated using our wide wavelength visible-NIR methods. The best models had prediction accuracy of about 0.1% over the FR concentration range that was typically 0–20% (Figure 14). The other FRs can be handled using other spectroscopies applied with the same instrumental methodology and software tools.

4.2 X-Ray Fluorescence and Optical Emission Spectroscopy

A complementary addition to molecular analysis is elemental analysis which provides information not only on the presence of halogens but also on other elements that are relevant to other QA metrics and also to good product stewardship. We chose to include commercially available portable instruments to align with the portability of TRANSPEC, a hand-held XRF instrument (Innov-X Inspector™) and an OES instrument (IoSys SSS3 Sliding Spark spectrometer (SSS)). Bromine content was measured for some FR/plastic combinations. The resolution of the XRF was typically better than 0.1% (more accurate and repeatable than the Sliding Spark instrument), but the overall results were in general agreement. The SSS is a viable cheaper option but limited to analysis of halogen content only, and requires regular cleaning. The

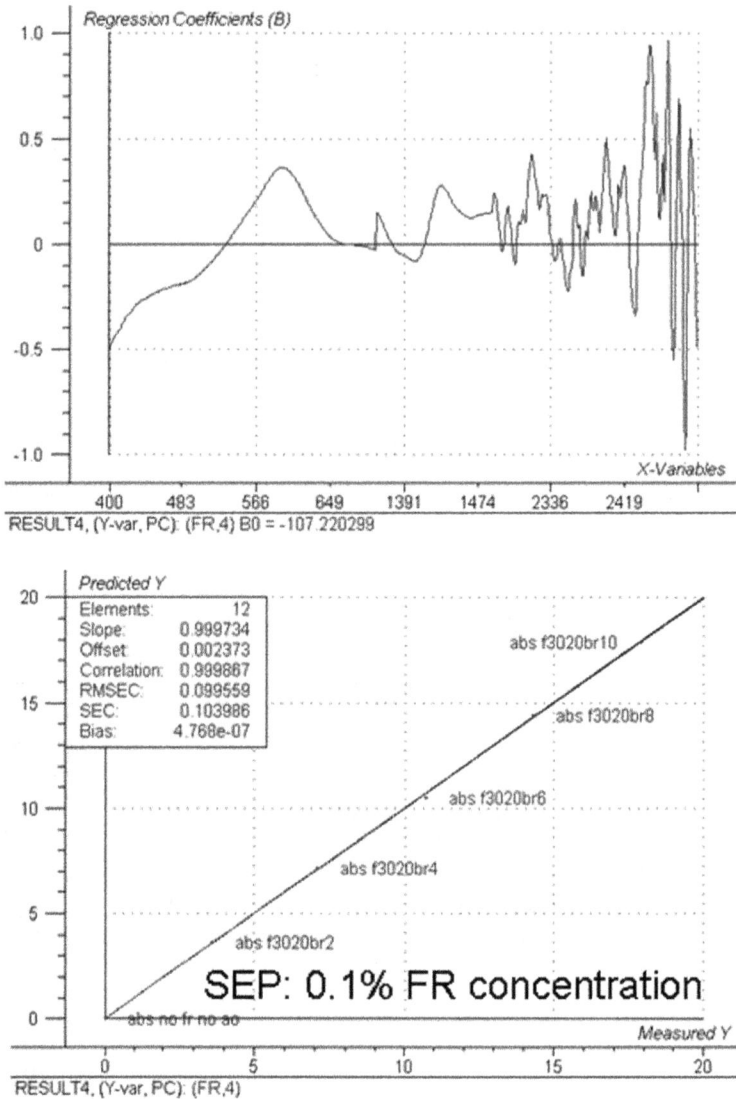

Figure 14   Improved model for ABS and epoxy FR.

XRF instrument was re-calibrated during the measurements. Some other elements were investigated with the XRF: levels of antimony appeared much as expected but significant levels of chlorine, calcium, barium and titanium were also detected. Chlorine levels in ABS were found to be about 1.3% in the case of 2% Br and about 4.5% in the case of 10% Br. Calcium levels were found to range between 0.8 and 4.6%. In the case of 10% Br, barium levels appeared at about 1.2% and titanium levels were about 0.4%. The levels of calcium and chlorine suggest that calcium chloride ($CaCl_2$) is an additional additive in the polymer when FR and antioxidant are present. $CaCl_2$ is an ionically bonded material, which explains the absence of detection in the subsequent Raman spectroscopy performed on the same samples (Section 4.3).

4.3 Infrared and Raman Spectroscopy

For plastics containing highly halogenated FRs it is necessary to use complementary spectroscopies whose data can be processed in the same way as the wide-wavelength (VIS-NIR) data using our MVSA methods. Other techniques typically used would be NMR or GC-MS which would require laboratory-based equipment. Infrared and Raman spectroscopy of plastics is well established[15,23–29] but similar analysis of plastics including flame retardants and other additives is limited.[30] The measurements were taken on two Perkin-Elmer spectrometer systems: an infrared spectrometer (using attenuated total reflection) and a Raman spectrometer. The flame retardants can be detected by comparing the spectra of polymer with and without FR (after accounting for the antioxidant, as before). The bromine components, undetectable in the NIR, are now identified. Raman was particularly useful in detecting the low-wavenumber carbon-bromine (C-Br) deformation vibrations accurately ($200–600\,\mathrm{cm^{-1}}$). Infrared provided some unique fundamental bond-stretching bands in the region $600–800\,\mathrm{cm^{-1}}$ and additional overtone information in the region $1300–1400\,\mathrm{cm^{-1}}$. For the purpose of reliable peak identification, Raman spectroscopy was also applied to the flame retardant and antioxidant powders separately, since the spectral features can be quite small compared with the main spectral peaks relating to the host plastic. However, these small differences are sufficient for our software to generate good prediction models. The spectral range in which the C-Br stretching vibrations are seen to occur is in good agreement with the available literature, and clearly relates to different modes of aromatic ring vibration that can be inferred from Raman spectra of simple aromatics such as benzene[31] and benzene derivatives.[32] PLS models were created using these Raman spectra and known concentrations of FR, and these indicate that FR content can be predicted using this method to an accuracy of better than 0.2%. Other peaks that occur in the low-wavenumber region are due to antimony trioxide, which is commonly present in plastics that contain brominated FRs. Raman spectroscopy of a similar plastic sample containing the antioxidant only shows four strong peaks in the $100–500\,\mathrm{cm^{-1}}$ range (Figure 15). The vibrations are referred to as v followed by a subscript number representing the type of vibration that is inferred from the literature.[33]

In addition to the additives, there is also a wealth of spectral information relating to the polymer content of the plastics that can be observed in these Raman spectra, much of which is well separated from the spectral features relating to additives. Ideally Raman spectroscopy could serve as a useful alternative method to NIR spectroscopy for polymer identification and

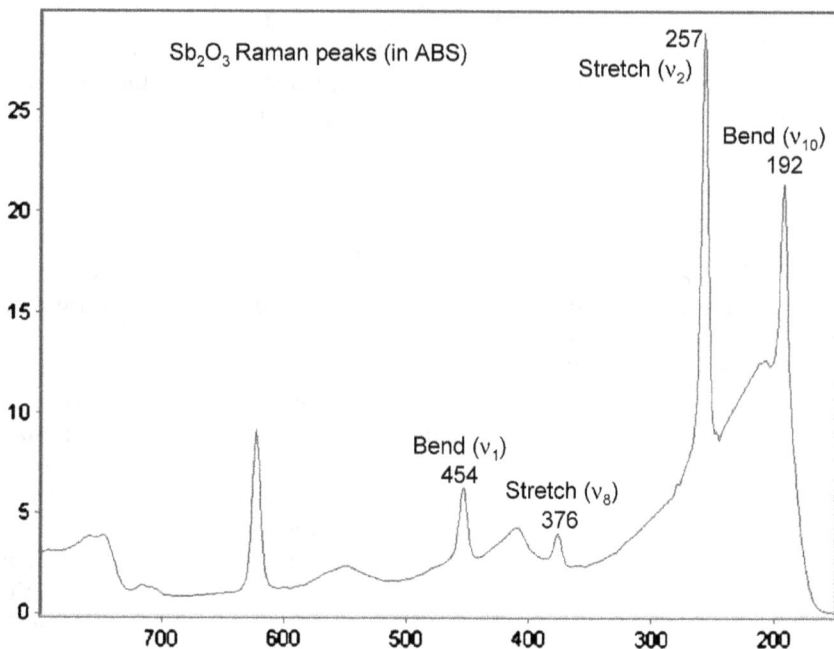

Figure 15 Raman spectrum of plastic showing antioxidant peaks.

condition assessment. We have investigated the use of commercially available hand-held Raman instruments but currently these are limited in the range and resolution that would be needed for reliable identification of these additives in plastics. At this stage the Raman measurements acquired for this purpose are limited to a laboratory environment only, but development of a hand-held device is presently being explored. Raman spectroscopy is identified as the preferred method over infrared ATR for measuring the FR powders separately, due to the clearer signals obtained in the low-wavenumber region in which the fundamental vibrations of C-Br and Sb-O bonds are found (100–700 cm^{-1}). However, in the case of black plastics, infrared would be the preferred method. Raman spectroscopy can provide a high degree of discrimination between highly halogenated and inorganic FRs, as seen from the examples shown in Figure 16. Limited previous Raman work on tetrabromobenzene[34] and other halogen-substituted aromatics[35,36] is useful for reference in assigning the C-Br vibrations in these FRs to the spectral features observed. There are significant differences found in these Raman spectra due to the different molecular groups involving bromine, and also other groups that are attached. Some combinations are more easily discriminated than others. Close examination of the spectral peaks in the 100–500 cm^{-1} range highlights the differences in bromine group deformations between the FRs.

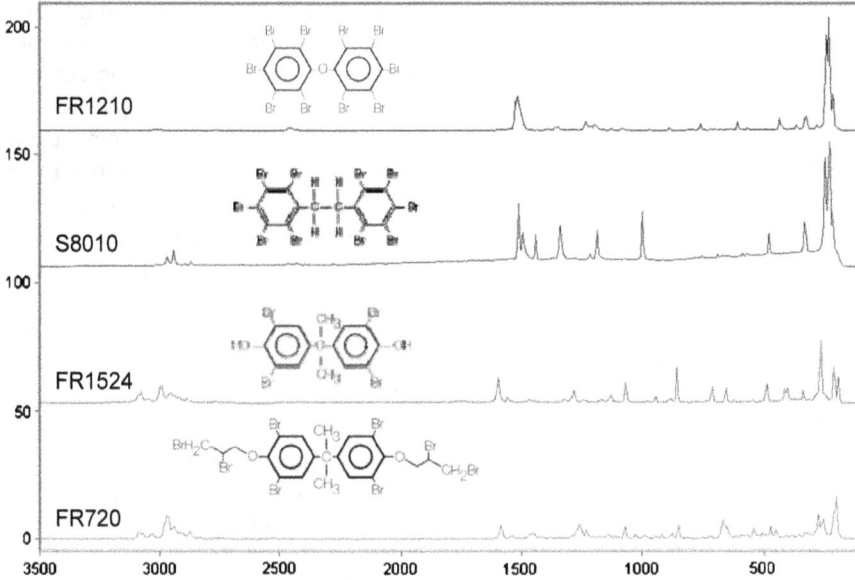

Figure 16 Raman spectra of FRs.

5 Conclusions

The rapid analysis and qualification of polymers is in many cases well suited to remote spectroscopic analysis. Although there are several methods for discrimination in polymers, molecular spectroscopic methods enable identification of particular molecular species in polymers and additives, as well as tracing the subtle effects of ageing and degradation that can in turn be related to critical polymer properties. If the hierarchy of questions starts with identification and then proceeds to additive discrimination at a concentration level above 0.1% w/w and finally grade or physical/mechanical property determination, then the use of wide-wavelength visible-NIR spectroscopy offers a rapid, robust and cost-effective analysis. We have shown that wide-wavelength spectroscopy with multivariate statistical analysis is capable of rapidly differentiating polymer types, identifying and quantifying the flame retardant content of plastics in at least half of the cases investigated, and provides information on physical and mechanical properties of the material which can then be used to estimate polymer grades and/or level of degradation. In the case of the epoxy FRs and TBBA, the prediction of FR content using visible-NIR spectroscopy was very high, with an accuracy 0.1% w/w or better.

FR additives that contain little organic hydrogen content, particularly in the form of hydroxyl groups, do not provide discrimination in the NIR spectral region. Raman can access these FRs, although in some cases it can be subject to other interferences such as fluorescence. Our Raman measurements on the plastics containing highly brominated FRs have shown very promising results and further work is in progress to produce efficient predictive multiple-parameter models that incorporate polymer identification, FR content and grade information into a single measurement. In the case of black plastics, the reflectance over this wavelength range can be too low for measurement and standard Raman is inappropriate, but other spectroscopic methods can be applied, such as mid-infrared using appropriate surface-probing techniques, and micro-Raman spectroscopy. If only elemental information such as bromine content is required then XRF or OES are appropriate alternative methods to use.

In these measurements, features which are difficult to distinguish by eye are readily discriminated by the family of multivariate statistical methods that utilise either the entire spectral data set, or certainly large parts of it. These methods are useful as they can be made relatively robust towards gross data changes that have little to do with samples, such as roughness and orientation, and provide a quantitative measure of the accuracy of property prediction. These techniques can be used to continually expand the existing data sets and models to include more plastic types and grades, and other types of flame retardants.

References

1. Directive 2002/96/EC of the European Parliament and of the Council of 27 January 2003 on Waste Electrical and Electronic Equipment (WEEE).
2. K. Freegard, G. Tan and R. Morton, *Develop a Process to Separate Brominated Flame Retardants from WEEE Polymers*, Project PLA-037, Final Report, The Waste & Resources Action Programme (WRAP), 2006.
3. X. Qu, A. S. Williams and E. R. Grant, *IEEE Trans. Electronics Packaging Manufacturing*, 2006, **29**, 25.
4. P. J. Baird, H. Herman, G. C. Stevens and P. N. Jarman, *IEEE Trans. Dielec. Electr. Insul.*, 2005, **13**, 309–318.
5. H. Martens and T. Naes, *Multivariate Calibration,* Wiley, New York, 1993.
6. P. J. Baird, H. Herman and G. C. Stevens, *IEEE Trans. Dielec. Electr. Insul.*, submitted 2007.
7. G. C. Stevens, H. Herman and P. J. Baird, *Polymers in Electronics 2007*, Smithers Rapra, Munich, 2007, No. 16.
8. P. Baird, I, Finberg, P. Georlette, H. Herman, W. Mortimore and G. Stevens, *Flame Retardants 2008*, Interscience Communications, London, 2008, 205.
9. G. C. Stevens, H. Herman and P. J. Baird, International Patent Application WO 2006/032910 A1, 2006.
10. P. J. Potts, A. T. Ellis, P. Kregsamer, C. Streli, M. West and P. Wobrauschek, *J. Anal. At. Spectrom.*, 1999, **14**, 1773–1799.
11. A. Golloch and T. Seidel, *Fresenius J. Anal. Chem.*, 1994, **349**, 32–35.
12. G. Zachmann, *J. Mol. Struct.*, 1995, **348**, 453–456.

13. F. Kowol, M. Oleimeulen, H. Freitag and T. Huth-Fehre, *J. Near Infrared Spectrosc.*, 1998, **6**, A149–A151.
14. A. Sadezky, H. Muckenhuber, H. Grothe, R. Niessner and U. Poschl, *Carbon*, 2005, **43**, 1731–1742.
15. S. R. Shield and G. N. Ghebremeskel, *J. Appl. Polym. Sci.*, 2003, **88**, 1653–1658.
16. C. Schade, W. Heckmann, S. Borchert and H. W. Siesler, *Polym. Eng. & Sci.*, 2006, **43**, 381–383.
17. D. L. Snavely and C. Angevine, *J. Polym. Sci. Part A*, 1996, **34**, 1669–1673.
18. T. Takeuchi, S. Tsuge and Y. Sugimura, *J. Polym. Sci. Part A-1*, 1968, **6**, 3415–3417.
19. S. C. de Araujo and Y. Kawano, *Polimeros: Ciencia e Tecnologia*, 2001, **11**, 213–221.
20. A. Boldizar and K. Moller, *Polym. Degr. & Stab.*, 2003, **81**, 359–366.
21. P. Davis, B. E. Tiganis and L. S. Burn, *Polym. Degr. & Stab.*, 2004, **84**, 233–242.
22. R. Ramani and C. Ranganathaiah, *Polym. Degr. & Stab.*, 2000, **69**, 347–354.
23. J. G. Bokria and S. Schlick, *Polymer*, 2002, **43**, 3239–3246.
24. P. A. M. Steeman, R. J. Meier, A. Simon and J. Gast, *Polymer*, 1997, **38**, 5455–5462.
25. D. Stevanovic, A. Lowe, S. Kalyanasundaram, P.-Y. B. Jar and V. Otieno-Alego, *Polymer*, 2002, **43**, 4503–4514.
26. S.-N. Lee, V. Stolarski, A. Letton and J. Laane, *J. Mol. Struct.*, 2000, **521**, 19–23.
27. C. Y. Liang and S. Krimm, *J. Polym. Sci.*, 1958, **27**, 241–254.
28. C. Y. Liang and S. Krimm, *J. Polym. Sci.*, 1958, **31**, 513–522.
29. J. Guilment and L. Bokobza, *Vib. Spectrosc.*, 2001, **26**, 133–149.
30. S. Kikuchi, K. Kawauchi, S. Ooki, M. Kurosawa, H. Honjho and T. Yagishita, *Analytical Sciences*, 2004, **20**, 1111–1112.
31. G. Herzberg, *Molecular Spectra and Molecular Structure,* Van Nostrand, Toronto, New York and London, 1945.
32. G. Varsanyi, *Vibrational Spectra of Benzene Derivatives,* Academic Press, New York and London, 1969.
33. S. J. Gilliam, J. O. Jensen, A. Banerjee, D. Zeroka, S. J. Kirkby and C. N. Merrow, *Spectrochim. Acta Part A*, 2004, **60**, 425–434.
34. K. M. White and C. J. Eckhardt, *J. Chem. Phys.*, 1998, **109**, 208–213.
35. P. V. R. Rao and G. R. Rao, *Spectrochim. Acta Part A*, 2002, **58**, 3039–3065.
36. N. Sundaraganesan, H. Saleem and S. Mohan, *Spectrochim. Acta Part A*, 2003, **59**, 1113–1118.

Subject Index